supporting expeditionary aerospace forces

AF709629

An Operational Architecture for Combat Support Execution Planning and Control

James Leftwich

Robert Tripp

Amanda Geller

Patrick Mills

Tom LaTourrette

C. Robert Roll, Jr.

Cauley von Hoffman

David Johansen

RAND

Project AIR FORCE

Prepared for the United States Air Force

The research reported here was sponsored by the United States Air Force under Contract F49642-01-C-0003. Further information may be obtained from the Strategic Planning Division, Directorate of Plans, Hq USAF.

Library of Congress Cataloging-in-Publication Data

Supporting expeditionary aerospace forces : an operational architecture for combat support execution planning and control / James Leftwich ... [et al.].
p. cm.
"MR-1536."
Includes bibliographical references.
ISBN 0-8330-3169-4
1. United States. Air Force. 2. Command and control systems—United States. 3. Operational art (Military science) 4. Air warfare. I. Leftwich, James, 1964–

UG633 .S855 2002
358.4'133041—dc21

2002067905

RAND is a nonprofit institution that helps improve policy and decisionmaking through research and analysis. RAND® is a registered trademark. RAND's publications do not necessarily reflect the opinions or policies of its research sponsors.

Cover design by Barbara Angell Caslon

Published 2002 by RAND
1700 Main Street, P.O. Box 2138, Santa Monica, CA 90407-2138
1200 South Hayes Street, Arlington, VA 22202-5050
201 North Craig Street, Suite 202, Pittsburgh, PA 15213
RAND URL: http://www.rand.org/
To order RAND documents or to obtain additional information, contact Distribution Services: Telephone: (310) 451-7002; Fax: (310) 451-6915; Email: order@rand.org

PREFACE

This report presents concepts for guiding development of an Air Force combat support (CS) execution planning and control operational architecture that meets the needs of the Expeditionary Aerospace Force (EAF). These concepts incorporate evolving practices; information from interviews with Air Force personnel; lessons from the Air War Over Serbia (AWOS), Operation Enduring Freedom (OEF), Operation Noble Eagle (ONE); and results of the authors' analysis of the current architecture for command and control of CS.

During the last few years, RAND has been defining the elements of a future Agile Combat Support (ACS) system that could help achieve AEF operational goals. The AEF operational goals are to

- Select and tailor force packages quickly to meet operational scenarios
- Deploy large and small force packages quickly
- Employ immediately with the capability to lay down firepower
- Shift smoothly to sustainment operations
- Deal quickly with changes to the campaign
- Allocate scarce resources to where they are needed most.

These goals place significant demands on the CS system, which must

- Estimate support requirements for alternative force packages, assess their feasibility, and propose alternative operational and support plans
- Estimate operational capabilities of beddown facilities and other combat support resources
- Configure the distribution network to meet employment and resupply needs
- Execute support plans and monitor support and operational performance
- Assess the effects of resource allocation options and prioritize allocations to users
- Signal when plans are out of control and support get-well analyses.

This study is one of a series of RAND publications that address ACS issues in implementing the EAF. Other reports in the series include the following:

- *Supporting Expeditionary Aerospace Forces: An Integrated Strategic Agile Combat Support Planning Framework,* Robert S. Tripp et al. (MR-1056-AF). This report describes an integrated ACS planning framework that can be used to evaluate support options on a continuing basis, particularly as technology, force structure, and threats change.
- *Supporting Expeditionary Aerospace Forces: New Agile Combat Support Postures,* Lionel Galway et al. (MR-1075-AF). This report describes how alternative resourcing of forward operating locations (FOLs) can support employment time lines for future AEF operations. It finds that rapid employment for combat requires some prepositioning of resources at FOLs.
- *Supporting Expeditionary Aerospace Forces: An Analysis of F-15 Avionics Options,* Eric Peltz et al. (MR-1174-AF). This report examines alternatives for meeting F-15 avionics maintenance requirements across a range of likely scenarios. The authors evaluate investments for new F-15 avionics intermediate-maintenance ship test equipment against several support options, including deploying maintenance capabilities with units, performing maintenance at forward support locations (FSLs), and performing all maintenance at the home station for deployment units.
- *Supporting Expeditionary Aerospace Forces: A Concept for Evolving the Agile Combat Support/Mobility System of the Future,* Robert S. Tripp et al. (MR-1179-AF). This report describes the vision for the ACS system of the future based on individual commodity study results.
- *Supporting Expeditionary Aerospace Forces: Expanded Analysis of LANTIRN Options,* Amatzia Feinberg et al. (MR-1225-AF). This report examines alternatives for meeting Low-Altitude Navigation and Targeting Infrared for Night (LANTIRN) support requirements for AEF operations. The authors evaluate investments for new LANTIRN test equipment against several support options, including deploying maintenance capabilities with units, performing maintenance at FSLs, and performing all maintenance at continental United States (CONUS) support hubs for deploying units.
- *Supporting Expeditionary Aerospace Forces: Lessons From the Air War Over Serbia,* Amatzia Feinberg et al. (MR-1263-AF). This report describes how the Air Force's ad hoc implementation of many elements of an expeditionary ACS structure to support the air war over Serbia offered opportunities to assess how well these elements actually support combat operations and what the results imply for the configuration of the Air Force ACS structure. The findings support the efficacy of the emerging expeditionary ACS structural framework and the associated but still-evolving Air Force support strategies.
- *Supporting Expeditionary Aerospace Forces: Alternatives for Jet Engine Intermediate Maintenance,* Mahyar A. Amouzegar et al. (MR-1431-AF). This report documents work on alternative concepts for Jet Engine Intermediate

Maintenance (JEIM) to determine whether peacetime and wartime jet engine maintenance is better performed by JEIM shops located with the aircraft or by organizations operating in a centralized facility.

- *Supporting Expeditionary Aerospace Forces: Forward Support Location Options,* Tom LaTourrette et al. (MR-1497-AF). This report assesses location options for intermediate-level maintenance of fighter aircraft. It identifies feasible sites that meet operational requirements for potential expeditionary operations and derives estimates of the investment and operating requirements and costs needed to implement a forward support location system. Candidate locations must be able to supply forward operating locations, have low wartime vulnerability, and be accessible for future U.S. use. (Limited distribution; not for public release.)

The research in this report was conducted in the Resource Management Program of Project AIR FORCE and was sponsored by the Air Force Deputy Chief of Staff for Installations and Logistics (AF/IL).

PROJECT AIR FORCE

Project AIR FORCE, a division of RAND, is the Air Force federally funded research and development center (FFRDC) for studies and analysis. It provides the Air Force with independent analyses of policy alternatives affecting the development, employment, combat readiness, and support of current and future aerospace forces. Research is performed in four programs: Aerospace Force Development; Manpower, Personnel, and Training; Resource Management; and Strategy and Doctrine.

CONTENTS

FIGURES

TABLES

SUMMARY

INTRODUCTION AND MOTIVATION

To be able to execute the full spectrum of aerospace operations, the United States Air Force has transitioned to an Aerospace Expeditionary Force (AEF).[1] Much of the discussion about the AEF concept has focused on changes in the way the Air Force is organized and provides forces to joint-service force commanders. The AEF construct concerns rapidly deploying, employing, and sustaining aerospace power around the globe, from a force structure that is predominantly located within the Continental United States (CONUS). These AEF global force projection goals present significant challenges to the current combat support (CS) structure. The AEF's requirement to respond quickly means that force and support packages must be tailored quickly to meet the operational needs of the specific contingency. The deployment and sustainment of CS resources must be coordinated to arrive at forward operating locations (FOLs) so that initial and sustained operations can take place without interruption. Most of the resources needed to support operations (munitions, housekeeping, and so forth) are not part of the deploying units. Scarce resources must be allocated to units with the highest priorities, often from different regions of the world. Thus, initiating and sustaining AEF operations require planning and control of a global network of CS resources from organic and industrial sources.[2]

AGILE COMBAT SUPPORT COMMAND AND CONTROL

This report presents concepts for guiding the development of a CS command and control operational architecture for the Aerospace Expeditionary Force. The concepts were developed from an analysis of AEF doctrinal changes, evolving

---

[1]When first introduced, the term EAF was used to describe the concept of employing Air Force forces rapidly, anywhere in the world, in predefined force packages called AEFs. The terms have since evolved and the Air Force now uses the term AEF to describe both the concept and force packages. Whereas previous RAND reports in the Supporting Expeditionary Aerospace Forces series refer to EAFs, we now use the term AEF to maintain consistency with Air Force usage.

[2]Previous RAND analyses offer recommendations for such an infrastructure, which would include forward operating locations from which missions would be flown and forward support locations/CONUS support locations for regional repair and storage facilities, a transportation system for distribution, and a combat support command and control system. See Tripp et al., *Supporting Expeditionary Aerospace Forces: A Concept for Evolving the Agile Combat Support/Mobility System of the Future*, RAND, MR-1179, 2000.

practices, Joint Universal Lessons Learned (JULLs) from exercises and experimentation, information from Air Force personnel, lessons from the Air War Over Serbia (AWOS), preliminary analysis of Operation Enduring Freedom (OEF) and Operation Noble Eagle (ONE), and results of our analysis of the current CS Command and Control (C2) operational architecture.

DEFINING CS EXECUTION PLANNING AND CONTROL AND OPERATIONAL ARCHITECTURE

Joint-service and Air Force doctrine defines C2 as the exercise of authority and direction by a properly designated commander over assigned and attached forces in the accomplishment of the mission.[3] Specifically, C2 includes the battlespace management processes of planning, directing, coordinating, and controlling forces and operations. It requires the integration of the systems, procedures, organizational structures, personnel, equipment, facilities, information, and communications that enable a commander to exercise command and control across the range of military operations.[4] In a narrow sense, this definition, because it deals with battlespace management, includes C2 functions with respect to the operational and tactical levels of warfare. We build on this definition of C2 and define CS execution planning and control to include the functions of planning, directing, coordinating, and controlling CS resources to meet operational objectives.[5] An operational architecture, by definition, describes the tasks, operational elements, and information flows required to accomplish or support a Department of Defense (DoD) function or military operation. It defines the types of information exchanged, the frequency of exchange, which tasks and activities are supported by the information exchanges, and the nature of information exchanges in sufficient detail to ascertain specific interoperability requirements.[6] For our study, we use these definitions, applied to Air Force CS activities, to identify and describe the processes involved in CS execution planning and control at each echelon and across each phase of operation.[7]

Our study defines and analyzes the current doctrinal CSC2 (*AS-IS*) architecture, identifies changes needed in the *AS-IS* architecture to realize AEF operational goals

[3]Joint Pub 1-02, *DoD Dictionary of Military and Associated Terms*, April 12, 2001.

[4]U.S. Air Force, *Air Force Basic Doctrine*, Air Force Doctrine Document 1 (AFDD-1), September 1, 1997.

[5]Although our work here primarily discusses the operational and tactical levels of warfare, we believe that the CS execution planning and control definition includes the strategic level as well—e.g., over the Program Objective Memorandum (POM) process in which CS plans are assessed, monitored, and controlled.

[6]Department of Defense, *C4ISR Framework Document Version 2.0*, December 18, 1997. The command, control, communications, computers, intelligence, surveillance, and reconnaissance (C4ISR) framework is intended to ensure that the architectures developed by geographic and functional unified commands, military services, and defense agencies interrelate between and among the organizations' operational, systems, and technical architecture views, and are comparable and integrated across joint-service and multinational organizational boundaries.

[7]Rather than view the results of this study as a combat support command and control (CSC2) operational architecture, which would promote the concept of a stovepiped, non-integrated architecture, we address CS execution planning and control processes in the context of the larger Air Force C2 architecture.

and correct deficiencies identified during recent contingencies, and sets forth concepts in some detail for the future (*TO-BE*) architecture.

CSC2 *AS-IS* SHORTFALLS AND RECOMMENDATIONS TO MEET THE *TO-BE* ARCHITECTURE

Our analysis of the Air Force's CS execution planning and control process revealed important shortfalls in the *AS-IS* architecture. These shortfalls can be grouped into four categories:

- Poor integration of CS input into operational planning
- Absence of feedback loops and the ability to reconfigure the CS infrastructure dynamically
- Poor coordination of CS activities with the joint-service community
- Absence of resource allocation/prioritization mechanisms across competing theaters.

We propose a *TO-BE* CS execution planning and control architecture system that would enable the Air Force to meet its AEF operational goals. The architecture would enable the CS community to quickly estimate support requirements for force package options and assess the feasibility of operational and support plans. The architecture would permit quick determination of beddown needs and capabilities, facilitate rapid Time Phased Force and Deployment Data (TPFDD) development, and support development and configuration of a theater distribution network to meet Air Force employment timelines and resupply needs. The *TO-BE* architecture would facilitate development of resupply plans and monitor performance, determine impacts of allocating scarce resources to various combatant commanders, indicate when CS performance deviates from desired states, and facilitate the development and implementation of "get-well" plans.

Finally, this report offers recommendations to help the Air Force CS community move from the current architecture to the future concept we describe. We recommend:

- Summarizing and clarifying Air Force CS doctrine and policy. The objectives and functions of execution planning and control must be recognized and codified in doctrine. The functions of concurrent development of plans among operators and CS personnel, assessment of plan feasibility, use of feedback loops to monitor CS performance against plans, and development of get-well planning need to be articulated and better understood.
- Creating standing CS organizations to conduct execution planning and control. The Air Force has supported one contingency after another for the last decade. Standing (permanent) organizations are needed to conduct CS functions and reduce turbulence and problems associated with the transition from supporting one contingency to reshaping support processes to meet the needs of another contingency.

- Training operations and CS personnel on each other's C2 roles. Understanding each other's responsibilities and methods can facilitate incorporation of both aspects into operational plans.
- Fielding appropriate information system and decision support tools to translate CS resource levels and processes into operational capabilities or effects. This will improve understanding of CS constraints or value for an operational planning option.

CONCLUSION

The strategic and operational environment and the AEF concept that addresses it present significant challenges to the current CS structure. To meet AEF stated objectives, the CS community is reexamining its current support system. Correcting deficiencies in CS execution planning and control as identified in this report is integral to the success of this effort.

ACKNOWLEDGMENTS

Numerous persons inside and outside of the Air Force assisted and supported our work. We thank Lieutenant General Michael Zettler, Deputy Chief of Staff, Installations and Logistics (AF/IL), for sponsoring this effort. General Zettler also sponsored some of our earlier research on the Aerospace Expeditionary Force. At the Air Staff, we are especially grateful to Sue O'Neal (AF/ILX) and Grover Dunn (AF/ILM). They took a personal interest in the project and participated in frequent update briefings to provide senior leader guidance in developing the study. We also thank Brigadier General Robert Mansfield (AF/ILI), Brigadier General Theresa Peterson (AF/ILT), Brigadier General Patrick Burns (ACC/CE), Brigadier General Robert Elder (CENTAF/CV), and their staffs for their support and critique of this work.

We are also extremely grateful to the many individuals who contributed their time and knowledge during our visits to Air Force installations and organizations. We wish to thank the many people at Air Combat Command; Pacific Air Forces; United States Pacific Command; Central Command Air Forces; United States Air Forces Europe; the Aerospace Command, Control, Intelligence, Surveillance and Reconnaissance Center; the Joint Staff; and others who have helped us with this work. The individuals contributing their time to these interviews are listed in Appendix A.

Our research has been a team effort with the Air Force Logistics Management Agency (AFLMA), whose support has been critical to our work. We wish especially to thank Colonel Ronne Mercer (AFLMA/CC) and Lieutenant Colonel Mark McConnell (AFLMA/LGM) for their support.

Finally, we wish to thank our Air Staff project officer, Colonel Connie Morrow (AF/ILXS), for her encouragement and support. We also thank Mr. Dick Olsen (AF/ILXX) for his continuous support and exchange of ideas. At RAND, Robert Kerchner and Kenneth Evers made key contributions to the research reported here. Sally Sleeper and Eric Peltz provided critiques that improved the report. Special thanks to Gina Sandberg and Jeanne Heller for working many hours on preparing iterations of this document that led to the publishing of this report. Special thanks to Sandra Wade-Grusky for preparing the HTML flowchart and database.

ACRONYMS

ACC	Air Combat Command
ACO	Airspace Control Order
ACS	Agile Combat Support
AC2ISRC	Aerospace Command and Control, Intelligence, Surveillance, and Reconnaissance Center
AEF	Aerospace Expeditionary Force
AF/IL	Hq USAF Installations and Logistics
AFFOR	Air Force Forces
AFLMA	Air Force Logistics Management Agency
AFMC	Air Force Materiel Command
AFTTP	Air Force Tactics, Techniques, and Procedures
AGE	Aerospace Ground Equipment
AIS	Avionics Intermediate Shop
ALC	Air Logistics Center
AMD	Air Mobility Division
AMOCC	Air Mobility Control Center
AOC	Air Operations Center
AOR	Area of Responsibility
AOS	Air Operations Squadron
ATO	Air Tasking Order
AWOS	Air War Over Serbia
CAP	Crisis Action Planning

CAT	Crisis Action Team
CE	Civil Engineering
CIRF	Centralized Intermediate Repair Facility
COA	Course of Action
CONOP	Concept of Operations
CONUS	Continental United States
CRC	Contingency Response Cell
CS	Combat Support
CSC2	Combat Support Command and Control
CSL	CONUS Support Location
C2	Command and Control
C4ISR	Command, Control, Communications, Computers, Intelligence, Surveillance, and Reconnaissance
DIRMOBFOR	Director Mobility Forces
DSOE	Deployment Schedule of Events
EAF	Expeditionary Aerospace Force
FLOW	Focused Logistics Wargame
FMC	Fully Mission Capable
FMSE	Fuel Mobility Support Equipment
FOL	Forward Operating Location
FSL	Forward Support Location
GAMSS	Global Air Mobility Support System
GIC	Global Integration Center
ICP	Inventory Control Point
IS	Installation Support
JAC2C	Joint Aerospace C2 Course
JAOP	Joint Air Operations Plan
JEIM	Jet Engine Intermediate Maintenance
JFACC	Joint Forces Air Component Commander

JFC	Joint Forces Command
JIPTL	Joint Integrated Prioritized Target List
JMC	Joint Movement Center
JTF	Joint Task Force
LRC	Logistics Readiness Center
LRU	Line Replaceable Unit
LS	Logistics Support
MAAP	Master Air Attack Plan
MAJCOM	Major Command
MOE	Measure of Effectiveness
MTW	Major Theater War
NAF	Numbered Air Force
OEF	Operation Enduring Freedom
OJT	On-the-Job Training
ONE	Operation Noble Eagle
OPLAN	Operations Plan
OPT	Operations Planning Team
OSC	Operations Support Center
PACAF	Pacific Air Forces
POL	Petroleum, Oils, and Lubricants
POM	Program Objective Memorandum
RAT	Redeployment Assistance Team
RSP	Readiness Spares Package
RSS	Regional Supply Squadron
SORTS	Status of Resources and Training Systems
SOS	Source of Supply
SRU	Shop Replaceable Unit
TACC	Theater Airlift Control Center
TDS	Theater Distribution System

TPFDD	Time Phased Force and Deployment Data
USAFE	United States Air Forces Europe
USTRANSCOM	U.S. Transportation Command
UTASC	USAFE Theater Air Support Center
UTC	Unit Type Code
WRM	War Reserve Materiel

Chapter One

INTRODUCTION

COMBAT SUPPORT COMMAND AND CONTROL AS A COMPONENT OF AGILE COMBAT SUPPORT

During the past decade, the United States military has supported continuous deployments of forces around the world, often on very short notice and for prolonged duration, to meet the needs of a wide range of peacekeeping and humanitarian missions or major contingency operations. These deployments have come from a smaller force based closer to home. The pattern of varied and fast-breaking regional crises appears to be the model for the foreseeable future and has prompted the United States to reassess how it prepares, maintains, and employs its military forces.[1] In response to this operating environment, the Air Force has reorganized into an Aerospace Expeditionary Force (AEF).[2] In the AEF concept, the Air Force presents forces in multiple, self-contained packages that are equipped to provide integrated, sustained force anywhere in the world on very short notice. A major premise of the AEF concept is that forces that are primarily stationed in the Continental United States (CONUS) can be tailored rapidly, deployed quickly, employed immediately, and sustained indefinitely as a viable alternative to a permanent forward presence. This premise, however, reduces the margin for error and places an increased emphasis on combat support. Although the form and structure of the AEF continues to evolve, it is clear this concept will play a central role in the future U.S. Air Force.

These AEF global force projection goals present significant challenges to the current combat support (CS) system,[3] and the importance of command and control (C2) has been identified as a key component of the AEF Agile Combat Support (ACS) system

---

[1]Donald Rumsfeld, *Defense Strategy Review,* June 21, 2001; and Donald Rumsfeld, *Guidance and Terms of Reference for the 2001 Quadrennial Defense Review,* June 22, 2001.

[2]M. Ryan, "Air Expeditionary Forces," DoD press briefing, 1998. When first introduced, the term EAF was used to describe the concept of employing Air Force forces rapidly, anywhere in the world, in predefined force packages called AEFs. The terms have since evolved and the Air Force now uses the term AEF to describe both the concept and force packages. Whereas previous RAND reports in the Supporting Expeditionary Aerospace Forces series refer to EAFs, we now use the term AEF to maintain consistency with Air Force usage.

[3]Throughout this report, we use the word system in the general sense to mean a combination of facts, principles, methods, processes, and the like. We use the expression information system to refer specifically to a product designed to manage data.

that needs attention.[4] This report presents concepts for guiding the development of a CS execution planning and control operational architecture for the Aerospace Expeditionary Force. Within the Department of Defense (DoD), an operational architecture is a description of tasks, operational elements, and information flows required to accomplish or support a DoD function or military operation. It describes the operational elements, assigned tasks and activities, and information flows required to support the warfighter. It defines the types of information exchanged, the frequency of exchange, which tasks and activities are supported by the information exchanges, and the nature of information exchanges in sufficient detail to ascertain specific interoperability requirements.[5] For our study, we used these definitions, applied to Air Force CS activities, to identify and describe the processes involved in execution planning and control, at each echelon and across each phase of operations.[6]

OBJECTIVES OF CS EXECUTION PLANNING AND CONTROL

Joint and Air Force doctrine defines command and control as the exercise of authority and direction by a properly designated commander over assigned and attached forces in the accomplishment of the mission.[7] It includes the battlespace management process of planning, directing, coordinating, and controlling forces and operations. Command and control involves the integration of the systems, procedures, organizational structures, personnel, equipment, facilities, information, and communications that enable a commander to exercise C2 across the range of military operations.[8] We expand on this definition of C2, typically applied to battlespace management, and address the functions of planning, directing, coordinating, and controlling CS resources to meet operational objectives. In a narrow sense, this definition, because it deals with battlespace management, includes C2 functions with respect to the operational and tactical levels of warfare.[9]

---

[4]Research at RAND has focused on defining the vision and evaluating options for an ACS system that can meet AEF operational goals. See Galway et al., *Supporting Expeditionary Aerospace Forces: New Agile Combat Support Postures*, RAND, MR-1075-AF, 2000. Additional research has identified the importance of CSC2 within the AEF ACS system. See Tripp et al., *Supporting Expeditionary Aerospace Forces: An Integrated Strategic Agile Combat Support Planning Framework*, RAND, MR-1056-AF, 1999.

[5]Department of Defense, *C4ISR Framework Document Version 2.0*, December 18, 1997. The command, control, communications, and computing intelligence, surveillance, and reconnaissance (C4ISR) framework is intended to ensure that the architectures developed by geographic and functional unified commands, military services, and defense agencies interrelate between and among the organizations' operational, systems, and technical architecture views, and are comparable and integrated across joint-service and multinational organizational boundaries.

[6]Rather than view the results of this study as a combat support command and control (CSC2) operational architecture, which would promote the concept of a stovepiped, non-integrated architecture, we address CS execution planning and control processes in the context of the larger Air Force C2 architecture.

[7]Joint Pub 1-02, *DoD Dictionary of Military and Associated Terms*, April 12, 2001.

[8]U.S. Air Force, *Air Force Basic Doctrine*, Air Force Doctrine Document 1 (AFDD-1), September 1, 1997.

[9]Although our work in this report deals primarily with the operational and tactical levels of warfare, we take a wider view and believe that the CS execution planning and control definition includes the strategic level as well, e.g., over the Program Objective Memorandum (POM) process where CS plans need to be assessed, monitored, and controlled. Some may argue that planning is not part of the functions of CS, but we define it to include this function, which is consistent with AFDD-1.

AEF operational needs provide further insights for CSC2 requirements, as shown in Table 1.1. Rapidly tailoring force packages requires that the system begin to generate support requirements based on desired operational effects alone. Combat support planners must coordinate closely with operators to estimate suitable force packages before such decisions are finalized. Early generation of CS requirements will contribute substantially to Course of Action (COA) assessment, focusing efforts on feasible COAs early in the planning process.

Rapid deployment requires that the C2 for combat support system provide force beddown plans and assessments quickly. Again, assessments must begin before plans are finalized, and therefore the capabilities and status of all potentially relevant airfields must be available. In addition, the status of in-theater resources must be continuously updated and effectively communicated to facilitate rapid Time Phased Force and Deployment Data (TPFDD) development.

Quick employment and subsequent sustainment require that theater and global distribution, maintenance, and supply operations be rapidly configured and expanded, and that global prioritization and allocation of combat support resources be rapidly shifted to the area of interest. Effectively allocating scarce resources requires that the system monitor resources in all theaters and prioritize and allocate resources in accordance with global readiness.

Finally, the system needs to be self-monitoring during execution and able to adjust to changes in either CS performance or operational objectives.

Table 1.1

CSC2 Functionality Required to Meet AEF Operational Goals

AEF Operational Need	CSC2 Requirement
Rapidly tailor force packages to achieve desired operational effects	Estimate CS requirements for suitable force package options; assess feasibility of alternative operational and support plans
	Identify and preplan potential operating locations
Deploy rapidly	Determine FOL[a] beddown capabilities for force packages and facilitate rapid TPFDD development
Employ quickly	Configure distribution network rapidly to meet employment timelines and resupply needs
Shift to sustainment smoothly	Execute resupply plans and monitor performance
Allocate scarce resources to where they are needed most	Determine impacts of allocating scarce resources to various combatant commanders and prioritize allocations to users
Adapt to changes quickly	Indicate when CS performance deviates from desired state and facilitate development and implementation of get-well plans

[a]FOL = Forward Operating Location.

PROBLEMS REVEALED

The need for this level of CSC2 functionality, as well as further insights into the needs of the CSC2 system, was revealed in Air Force operations [Operation Noble Anvil (ONA)] in the Air War Over Serbia (AWOS). The lessons from and shortcomings in the present system in ONA provide useful insights for AEF CSC2 needs. The major lessons and corresponding CSC2 requirements are summarized in Table 1.2. Initial analysis of Operation Enduring Freedom (OEF) and Operation Noble Eagle (ONE) revealed many of the same shortcomings.

The transition to wartime CS operations in ONA was difficult, partly because of a departure from doctrine that shifted command from the Numbered Air Force (NAF) to the Major Command (MAJCOM) during operations. Although there were several reasons for this action,[10] shifting organizational responsibilities during conflict created problems, including attempting to use organizational relationships that did not exist day to day, delays in developing communications paths, and using people who may not have been trained for wartime jobs. These problems may have arisen even if the NAF had supported ONA because of the staff augmentation necessary to make the current doctrinal organization effective.[11] There is a need for standing (permanent) CS organizations to provide operational continuity and seamless transition through the spectrum of operations from peacetime to major theater warfare. The transition

Table 1.2

CSC2 Requirements Revealed by Lessons from Operation Noble Anvil

ONA Lesson	CSC2 Requirement
Slow and difficult transition from peacetime to wartime operations	Identify permanent organizations that will perform critical CS tasks continuously during peace and war
	Expand Air Force involvement in theater distribution system planning and execution
Poor interface between operations and CSC2	Include CS input in initial planning
	Translate CS information into operations capability
Inability to react quickly to changes in the operational plans	Provide real-time visibility of theater and global resources
	Rapidly reconfigure CS infrastructure
Insufficient and inadequate CSC2 policy/procedures, systems, training, and education	Develop and formalize doctrine/policy, systems, and training programs

[10]Feinberg et al., *Supporting Expeditionary Aerospace Forces: Lessons From the Air War Over Serbia*, RAND, MR-1263-AF, 2002.

[11]Hanser et al., *The Warfighting Capacity of Air Combat Command's Numbered Air Forces*, RAND, DB-297-AF, 2000.

was also hindered by confusion over responsibility for theater distribution management, leading to an ineffective theater distribution system.[12] Rapid configuration of both theater and global distribution systems is essential to meet AEF operational deployment and employment goals, further highlighting the need for these responsibilities to be clearly delineated between the services or for the Air Force to maintain the skills to develop and configure such a system.

In AWOS, the limited ability and opportunity for interaction between the CS and operations planners led to plans being developed with minimal CS input, resulting in excessive revamping and slow progress.[13] Limited communications links between operations planners and CS planners hindered interaction. The single CS person responsible for interaction in the operational planning group did not have a full depth of CS experience, information system links, or decision support tools to help facilitate interaction. This lack highlights the importance of formalized procedures for including CS factors in operational planning and execution and relating CS capabilities in operational terms.

The CS system was slow to react to changes in the air campaign.[14] It was slow to reconfigure the CS support infrastructure (to redirect materiel, adjust maintenance priorities, and alter distribution routes and modes) partly because personnel were inexperienced in these wartime functions. According to doctrine, the NAF was to exercise these responsibilities. MAJCOM personnel taking on these functions were not trained in many of them. Also, the Air Force took on some functions, such as planning the theater distribution system, that it may not have trained personnel for, assuming that the Joint Command would have the wartime responsibility.

Finally, ONA demonstrated that existing policies, procedures, and information systems are inadequate and that education and training are insufficient.[15] While good people compensated for these shortfalls, the deficiencies did result in additional time to determine what should be done. We emphasize the importance of defining the role of CS execution planning and control and of incorporating those activities into training and education.

In summary, the CS execution planning and control system must be able to continuously monitor CS resource levels and translate them into operational metrics; evaluate the resources needed to achieve operational goals, assess the feasibility of support options, and help to develop alternative plans; rapidly reconfigure the CS infrastructure to meet specific contingency scenario needs; employ commodity and process control metrics and process monitoring to regulate support processes; and adjust support activities during execution to optimize warfighter support.

---

[12]Feinberg et al., 2002.

[13]Feinberg et al., 2002.

[14]Feinberg et al., 2002.

[15]Feinberg et al., 2002.

DEVELOPING AN OPERATIONAL ARCHITECTURE FOR CS EXECUTION PLANNING AND CONTROL

Our objectives were to define and analyze the current (*AS-IS*) architecture, identify changes needed, and present concepts for a future (*TO-BE*) architecture for the Air Force to use as a point of departure. A CS execution planning and control operational architecture should concentrate on the decisions by Air Force CS organizations and the information flows supporting these decisions over the phases of operations. In this analysis, we focus on sortie production, base support, and decisions made by each organization during all phases[16] of operations.

Based on our analysis of the *AS-IS* and *TO-BE* architectures, we identify the shortfalls in the *AS-IS* system and the changes required to achieve the functionality of the *TO-BE* system. We highlight the roadblocks in meeting AEF operational goals. We then present concepts for guiding the development of the *TO-BE* CS process, including an example of how the CS execution planning and control system would operate in a small-scale conflict. After discussing the existing shortfalls and modifications proposed in doctrine and policy, organizations, training and education, and tools and systems to move to the *TO-BE*, we summarize our findings, recommendations, and steps needed to implement the architecture.

---

[16]Air Force and joint-service publications refer to phases of operations differently. For our analysis, we have used readiness, deployment, employment, sustainment, redeployment, and reconstitution to describe the phases.

Chapter Two

ANALYSIS APPROACH

The objective of our research is to develop a set of concepts and a draft CS execution planning and control operational architecture that can support the Air Force of the future. The research should provide a solid foundation for the Air Force to use in developing and refining its overarching C2 operational architecture. The approach we used is shown in Figure 2.1.

The first step was to define the expected CS execution planning and control functionality. Our starting point was the operational needs of the Aerospace Expeditionary Force and lessons from recent contingencies. To this initial set of requirements, we incorporated extensive input from discussions with subject matter experts and site visits to over 20 Air Force and joint-service organizations (see Appendix A). We also included insights from previous studies, such as U.S. Joint Forces Command's *Focused Logistics: Enabling Early Decisive Operations.*[1]

We also documented the current *AS-IS* CS operational architecture by reviewing Air Force and joint-service doctrine, manuals, instructions, and concepts of operations (CONOPs); and describing processes and organizational responsibilities derived from the expert interviews, analyses of lessons learned from the AWOS, recent

RANDMR1536-2.1

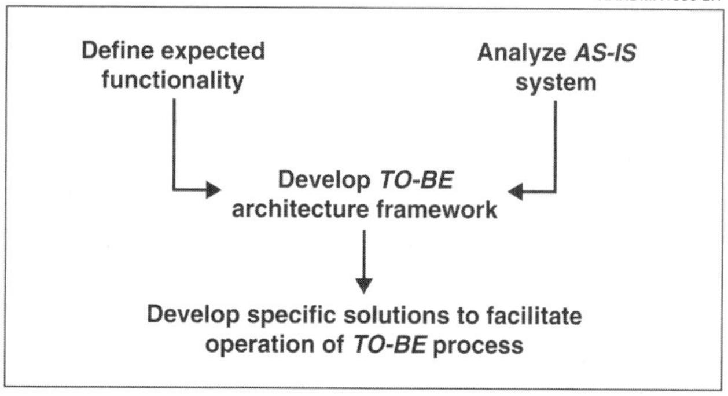

Figure 2.1—Analysis Approach

[1]U.S. Joint Forces Command, Concepts Division, *A White Paper for Focused Logistics: Enabling Early Decisive Operations*, October 1999.

contingencies, and insights from previous studies such as the base-lining effort conducted by Aerospace Command and Control, Intelligence, Surveillance, and Reconnaissance Center (AC2ISRC).[2]

Using the desired functional characteristics and analysis of the *AS-IS* architecture, we developed *TO-BE* concepts and an associated operational architecture. We present the architecture in a number of process maps, flow diagrams, and associated databases. The process maps have three levels of detail: a high level that shows the generalized concepts of CS execution planning and control that are applied across all phases of operations; a mid-level architecture that describes the concepts fairly closely; and a detailed architecture. The high- and mid-level process maps are described in Chapters Three and Four. The low-level *AS-IS* diagram, documented using flow-charting software, is contained in Appendix B. The low-level *TO-BE* diagram is shown in Appendix C. In addition to the process diagrams, the *TO-BE* operational architecture is documented in a database containing process activities and tasks in a hierarchical structure. The database includes information required to perform the tasks, the information sources, products produced by each activity and the recipients of the products. It identifies the organizational nodes responsible for the activities and tasks. The information in the database is consistent with the high-, mid-, and low-level process diagrams. The associated compact disk contains a complete set of the documentation described above.

The database that describes the information inputs for a given CS decision can be accessed by clicking on the C2 process box of interest in the HTML diagram. The *TO-BE* process diagram is hyper-linked to its more detailed database to show supporting tasks, information flows, and organizational node responsibilities. The HTML diagram also describes how one can navigate between the *TO-BE* diagram and the databases.

This framework allows one to drill down and follow how the general principles are used at lower and lower levels. This approach helps track how individual tasks align with the higher-level desired functionality, and also helps highlight redundancies and shortcomings in the *AS-IS* system.

The *AS-IS* architecture analysis was then compared to the AEF CS execution planning and control needs to identify *AS-IS* shortcomings and changes necessary for the *TO-BE* architecture. Shortcomings were broadly grouped according to the type of modification ("solution") that would address them. Solution themes are proposed to guide the development of more specific solutions. The categories of shortcomings and corresponding solution themes are shown in Table 2.1. For each category, we discuss several shortcomings and how each hinders efficient CS execution planning

---

[2]Aerospace Command and Control, Intelligence, Surveillance, and Reconnaissance Center, *USAF Command and Control CONOPs*, Vol. III, *Blue Order of Battle, Global Awareness for Expeditionary Aerospace Forces*, Langley Air Force Base, VA, July 7, 2000. While various CONOPs, doctrine and policy publications, and operating instructions mention CSC2, this is the first complete documentation of the operational architecture. Major sources from which this architecture was drawn are listed in the document's references.

Table 2.1

Summary of *AS-IS* Shortcoming Categories and Solution Themes

Shortcoming Category	Solution Theme
Doctrine/policy	Clarify objectives and functions
Organizations	Evolve to standing CS organizations to conduct C2 functions
Training and education	Adequately train, educate, and exercise CS personnel (e.g., train CS personnel on ops planning processes and ops personnel on CS planning processes)
Information systems/tools	Field appropriate capabilities

and control. We then propose solutions aimed at resolving the shortcomings. The solutions are designed to facilitate, enhance, and refocus the operation of the CS system to be in line with the desired functionality and *TO-BE* concepts.

Chapter Three

CSC2 *AS-IS* ARCHITECTURE: DESCRIPTION AND ANALYSIS

For many reasons, the Air Force does not have a defined and well-understood *AS-IS* C2 operational architecture for CS. First, in our interviews with informed service members, we noted the absence of a defined and documented baseline. Most responses were limited to personal experiences rather than well-documented processes and activities. Many interviewees identified shortfalls (described in Chapter Five) that are merely symptoms of the larger issue of not having a well-defined and documented operational architectural baseline. Second, responsibility for CS is fragmented, leading to planning and execution activities that are more decentralized than those of operations.[1] This decentralization of functions contributes to the variability in process descriptions and adds to the dilemma of who the operational planner should turn to for an overall view of combat support.[2]

We start with a process description of the *AS-IS* architecture that summarizes CS execution planning and control activities in several general steps. We examine the key tasks, operational elements, and information flows in the current system. The picture is Air Force-centric but will include joint-service entities and decisions as they pertain to Air Force CS activities. Finally, we discuss several process deficiencies, derived from our interviews and lessons learned from ONA,[3] that have hindered the ability of the CS system to support AEF operational goals.

---

[1]Col Bill McGill, AC2ISRC, Langley Air Force Base, indicated that "Combat support is done by committee, not by command."

[2]LGen Lance Smith, 7AF Commander, Osan Air Base, Korea, indicated that it is difficult to get a complete picture of combat support because there are so many players (e.g., force protection, logistics, civil engineering, and services). For this reason, it is also difficult to determine who the commander should go to to get the whole picture. MGen (S) Donald Wetekam (ACC/LG), Langley Air Force Base, indicated that Air Combat Command (ACC) co-located AFFOR A-4/7 Rear functions of civil engineering, services, and logistics to provide an integrated view of Operation Enduring Freedom. A senior combat support person could then speak for all these CS functions in planning for OEF with the forward echelon AFFOR staff. Throughout this report are references to a COMAFFOR and associated staff functions. Per Air Force Doctrine Document 2 (AFDD-2), the COMAFFOR is the Commander of Air Force Forces participating in a joint military operation. The COMAFFOR staff is organized into an Air Operations Center (AOC) that performs the operations tasks and an A-Staff that performs the support tasks. The A-staff is further organized along functional lines with the A-1 responsible for manpower and personnel, the A-2 responsible for intelligence, the A-3 responsible for operations (normally the director of the AOC), the A-4 responsible for combat support (logistics and installations), the A-5 responsible for planning, and the A-6 responsible for command, control, communications and computer systems. In some instances, the A-4 responsibilities are separated into an A-4 responsible for logistics and an A-7 responsible for installations.

[3]Feinberg et al., 2002.

AS-IS PROCESS MAPS AND DESCRIPTIONS

We have categorized the many activities and decisions of the CS execution planning and control system into six basic activities—high-level guidance, operational planning, CS tasks, assessment, execution, and evaluation.[4] Figure 3.1 provides a graphical representation of this condensed process.

The process is initiated by guidance that generates the demand for operational performance or output. The guidance generally provides the high-level objectives the operators will try to achieve. The operations plan will describe how forces will be used to achieve the operational objectives and often includes measures of effectiveness (MOEs) against which the iterations of the plan and its execution will be evaluated. During planning, operational effectiveness assessments are used to evaluate whether the plan will produce the intended results and lead to accomplishment of the objectives. Oftentimes, the operations plan will assume the availability of needed CS resources.

Once an operationally feasible plan has been developed, the CS community starts to prescribe tasks for supporting the plan. Combat support tasks flow from the planning outputs in the prescribed time frames.

When operations commence, CS tasks must keep pace with the dynamic operational requirements. Changes are executed in response to short-notice requests for support or to CS shortfalls.

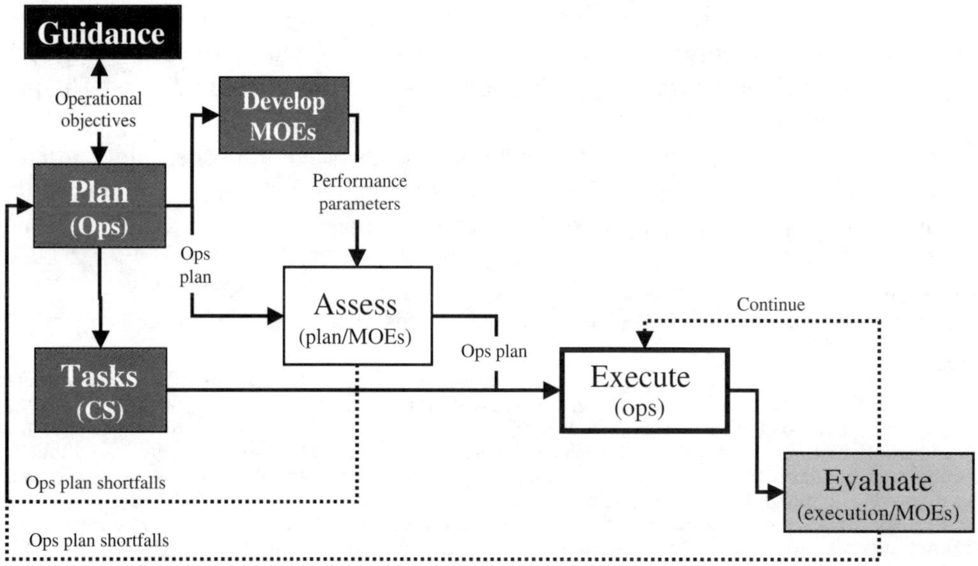

Figure 3.1—CSC2 High-Level *AS-IS* Process Description

---

[4]Although the six basic activities are not specifically discussed in doctrine or policy, they are generally described in AFDD-2, *Organization and Employment of Aerospace Power,* and related implementing instructions.

Combat operations effectiveness is evaluated daily; the operational assessment feedback loop is a critical input to planning the next day's sorties. Adjustments are made to the long- and short-range operations plans based on the effectiveness of each day's operations.

MORE DETAILED *AS-IS* ARCHITECTURE DESCRIPTION

We next highlight key tasks, operational elements, and information flows in the current CS execution planning and control operational architecture. Figure 3.2 shows at a lower level of detail the activities described above in the high-level process. Appendix B contains a detailed diagram of these activities along with the sources from which these activities were taken.

The Joint Chiefs of Staff (JCS) and theater combatant commanders continually monitor world situations and evaluate the need for military action. The activities associated with day-to-day operations are reflected in the first gray-shaded area in Figure 3.2. The Air Force focuses on training, equipping, and readying the force for combat. The Air Staff provides the service policy and guidance. *Force providing* MAJCOMs are responsible for maintaining the forces, while *support providing* MAJCOMs such as Air Force Materiel Command (AFMC) are responsible for maintaining sufficient support resources and processes. Both types of MAJCOM are responsible for ensuring that personnel are trained and equipment is maintained. Numbered Air Forces are identified by doctrine as the Air Force's "senior warfighting echelon."[5] They are responsible for developing host nation (HN) relationships and agreements, performing site surveys, and developing deliberate plans for their area of responsibility (AOR). Units are responsible for training their personnel, maintaining their equipment, and reporting status[6] up to their MAJCOM. Depots are responsible for maintaining materiel and processes needed to support peacetime and wartime operational requirements.

According to *AS-IS* doctrine, when a world situation induces the JCS to pursue a military COA, the JCS issues a warning order. The activities associated with strategy development in response to a crisis are reflected in the second gray-shaded area in Figure 3.2. The theater combatant commander creates a Joint Task Force (JTF) and appoints a JTF commander, who begins strategy development by planning a COA. If there is a standing JTF or sub-unified command, the combatant commander directs planning to begin. During strategy development, the NAF moves to its role as the Air

---

[5]AFDD-2.

[6]Unit preparedness status is accomplished through Status of Resources and Training System (SORTS) reports prescribed in Air Force Policy Directive 10-2, *Readiness*, and Air Force Instruction 10-201, *Status of Resources and Training System.*

RANDMR1536-3.2

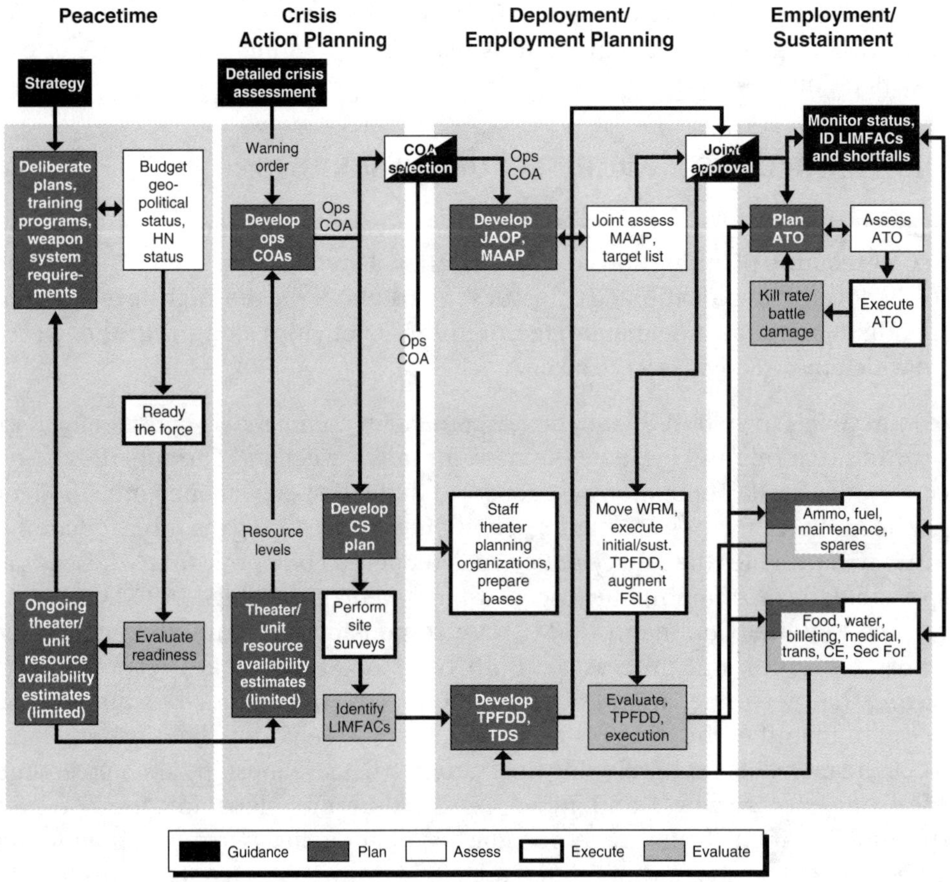

NOTES: JAOP = Joint Air Operations Plan; MAAP = Master Air Attack Plan; TDS = Theater Distribution System; FSL = Forward Support Location; ATO = Air Tasking Order; WRM = War Reserve Material; HN = Host Nation; LIMFACs = Limiting Factors

Figure 3.2—CSC2 Mid-Level *AS-IS* Process Description

Force Forces (AFFOR) staff[7] and is augmented to the appropriate level.[8] The AFFOR Battle Staff/Crisis Action Team (CAT) becomes the center of Air Force planning, monitoring, and decisionmaking during a contingency. The AFFOR personnel proceed with campaign planning (if the planning was initiated by a MAJCOM or special planning cell; otherwise the AFFOR staff initiates the operations planning and identifies logistics and installations support requirements) and survey potential beddown locations. During this stage, the AFFOR CAT receives general force requirements from the combatant commander/JTF and identifies specific combat and support ca-

[7]AFDD-2 outlines the doctrine. Each NAF has augmentation plans. For instance, for a major theater war (MTW) 7th Air Force receives 34 augmenters in A-4 functions from the 701st Combat Operations Squadron, and its 18 augmenters for A-7 (if established) functions from the Air National Guard S-Team. (Interviews with 7th Air Force, Osan, Korea.)

[8]Interviews with 32nd Air Operations Squadron (AOS), Ramstein Air Base, Germany, April 4, 2001.

pability requirements, designated by Unit Type Codes (UTCs). The AFFOR staff might have an initial plan, developed during peacetime, with which to work. Force selection and deployment timing information is documented in the Time Phased Force and Deployment List.[9] The TPFDD is developed by the AFFOR A-5, Director of Plans, in conjunction with combat support planners, who estimate CS requirements based on the operational plans provided and a series of logistics and installations support planning factors. Units that might be tasked are notified to prepare for possible deployment. The TPFDD is further refined and taskings are finalized as a COA is developed. The TPFDD then undergoes a validation review with the JTF and U.S. Transportation Command (USTRANSCOM), which is responsible for supporting strategic moves.[10] The planned force package (UTC) requirements are forwarded to the supporting commands, which, in turn, task specific units to provide military forces. Units are notified of their deployment schedule and await their deployment orders. Personnel in supporting commands monitor the deployment preparation and execution activities of their units and report limiting factors (LIMFACs) and shortfalls to the AFFOR and AF levels.

Once the COA is selected, the JCS issues an alert order and operational planning is further refined. The third gray-shaded area in Figure 3.2 reflects the activities associated with deployment and employment planning. Planners start with a high-level Joint Air Operations Plan (JAOP) that identifies the various air operations phases. During this planning activity, the air campaign plan can provide long-range (weeks to months) expectations of the demands on weapon systems. Once the campaign plan is developed, the operators begin the mid-term (typically 72–96 hours) planning called strategy, guidance, apportionment, and targeting (STRAT/GAT), during which they apportion weapon systems to general target sets in accordance with Joint Forces Air Component Commander (JFACC) guidance and provide some indication of the munitions types that will be used against the target sets. During the Master Air Attack Plan (MAAP) planning activity (48–72 hours), the operators start assigning mission sets to specific units.[11] The MAAP phase leads into the Air Tasking Order (ATO)/Air Coordination Order phase. During this phase, the operators assign specific times-over-target and standard conventional munitions loads. Operational planners determine daily mission tasking to support the combatant commander and JFACC objectives.

In parallel, or ahead of the detailed operational planning just discussed, the AFFOR A-4 should plan resupply and sustainment and estimate intratheater movements to the joint command that is responsible for planning and executing intratheater transportation. This plan is then executed by the service assigned as the executive agent for the Theater Distribution System (TDS). Concurrently, the JCS assigns movement priorities among theaters and services to balance strategic lift allocations among

---

[9]This is often referred to as Time Phased Force and Deployment Data (TPFDD), which is the data in the information system from which the Time Phased Force and Deployment List can be produced.

[10]A detailed description of the TPFDD development process is contained in Air Force Manual 10-401, *Operation Plan and Concept Plan Development and Implementation.*

[11]While we reflect a range of days and hours of planning provided by JAOP, STRAT/GAT, MAAP, and ATO planners, the number of days and hours can vary by scenario.

competing demands. USTRANSCOM subsequently develops transportation schedules for strategic lift of units deploying. After units receive their deployment orders, they arrive at their reception sites, prepare for sortie generation, and commence operations upon receipt of the ATO. In addition, the AFFOR installations support functional planners, determine base development, and build-up requirements for ramp space, utilities, billeting capacity, etc. to meet the timing for arrival of operational forces. The AFFOR monitors and assesses base development and reception and reports to the JTF. The supporting commands ready sustainment UTCs in preparation for ongoing operations.

When the JCS gives the execution order, units begin sortie production and execute their prepared ATOs. Employment, sustainment, and beddown activities are reflected in the right-most gray-shaded area in Figure 3.2. The results of the daily missions are fed back to the planners and are a key factor in determining future missions. This execution evaluation, using metrics such as kill rate and battle damage, is passed up through the AOC to the joint-service level. The AFFOR logistics planners receive logistics status reports and situation reports for each resource commodity and summarize them in a daily report for the AFFOR A-4, who in turn reports to the JTF J-4. The AFFOR installations support planners receive facilities, services, and infrastructure status reports and integrate them for the AFFOR and, in turn, for the JTF.

While ATO planning and execution continues, sustainment resources (food, water, ammo, fuel, spares, etc.) flow through the strategic and theater distribution system to FOLs and Forward Support Locations (FSLs) where they are needed. The Director, Mobility Forces (DIRMOBFOR), the JFACC's air mobility adviser, coordinates between transportation providers (Theater Airlift Control Center, Air Mobility Control Center, Air Mobility Division) and the services that require transportation. The Joint Movement Center (JMC) adjudicates competing demands for transportation resources and allocates daily flights for sustainment cargo. During execution, units with repair capability and Centralized Intermediate Repair Facilities (CIRF) perform maintenance to support sortie production. Base development and construction requirements flow to the Regional Wartime Construction Manager (RWCM), who coordinates them with the Joint Command for allocation to the services according to combatant commander priorities.

Redeployment, triggered by an order issued by the JCS, returns personnel and assets to their home bases. Following the order, joint-service and Air Force redeployment assistance teams are activated to assess redeployment. Once the JCS and combatant commander approve force removal, the AFFOR prepares the redeployment TPFDD. Upon receipt, units redeploy and assess resources.

ANALYSIS OF *AS-IS* PROCESS SHORTFALLS

Our analysis of the Air Force's CSC2 process revealed shortfalls in the *AS-IS* system that can be categorized as follows:

- Poor integration of CS (logistics and installations support) inputs into operational planning
- Absence of feedback loops and the ability to dynamically reconfigure the CS infrastructure
- Poor coordination of CS activities with the joint-service community
- Absence of resource allocation arbitration across competing services and theaters
- Inadequate understanding that combat support means not only logistics but installations support as well.

Poor Integration of CS Input into Operational Planning

The conventional roles of the operations and combat support communities often entail separate and relatively independent C2 activities. Operational plans may be developed without adequate regard to CS feasibility.[12] Figure 3.3 shows where this disconnect affects the planning and execution process. Early in the planning process, the strategy cell, consisting of A-3 and A-5 planners, recommends COAs to the JFACC. A-4 (and A-7 if established) representatives may be present during plan development but not be called upon to evaluate plan feasibility. Combat support personnel must then generate the appropriate resources to support a particular TPFDD or ATO. This serial approach can result in prolonged development of unsupportable plans and require major restructuring when CS factors are eventually brought to light, or result in unnecessary resource expenditures. An example of this occurred in ONA. One combat support person from munitions was involved in the development of the evolving COA.[13] As a result of the poor interface between CS personnel and operational planners, munitions personnel responsible for resupplying FOLs sometimes wrongly estimated the munitions requirements. Because resupply had to be taken in advance, on occasion quantities of munitions just sent to an FOL (e.g., Aviano Air Base) would have to be returned to the munitions storage area (e.g., Camp Darby) because of limited storage at the FOL and the estimated expenditures not materializing. This double movement is costly and can be avoided with better interface between operations planners and CS personnel.[14] In another case, operational planners chose the location for a potential beddown area without sufficient installation support planning input, resulting in the tent city having to be relocated both because of flooding and interference with explosive safety areas.

The traditional separation between the combat support and operational planning communities hinders effective integration. Most logisticians, for example, are not trained in and do not participate in air campaign planning. They therefore have little

---

[12]Col Ed Groeninger, PACAF 502/CC, March 8, 2001.

[13]Feinberg et al., 2002.

[14]Feinberg et al., 2002.

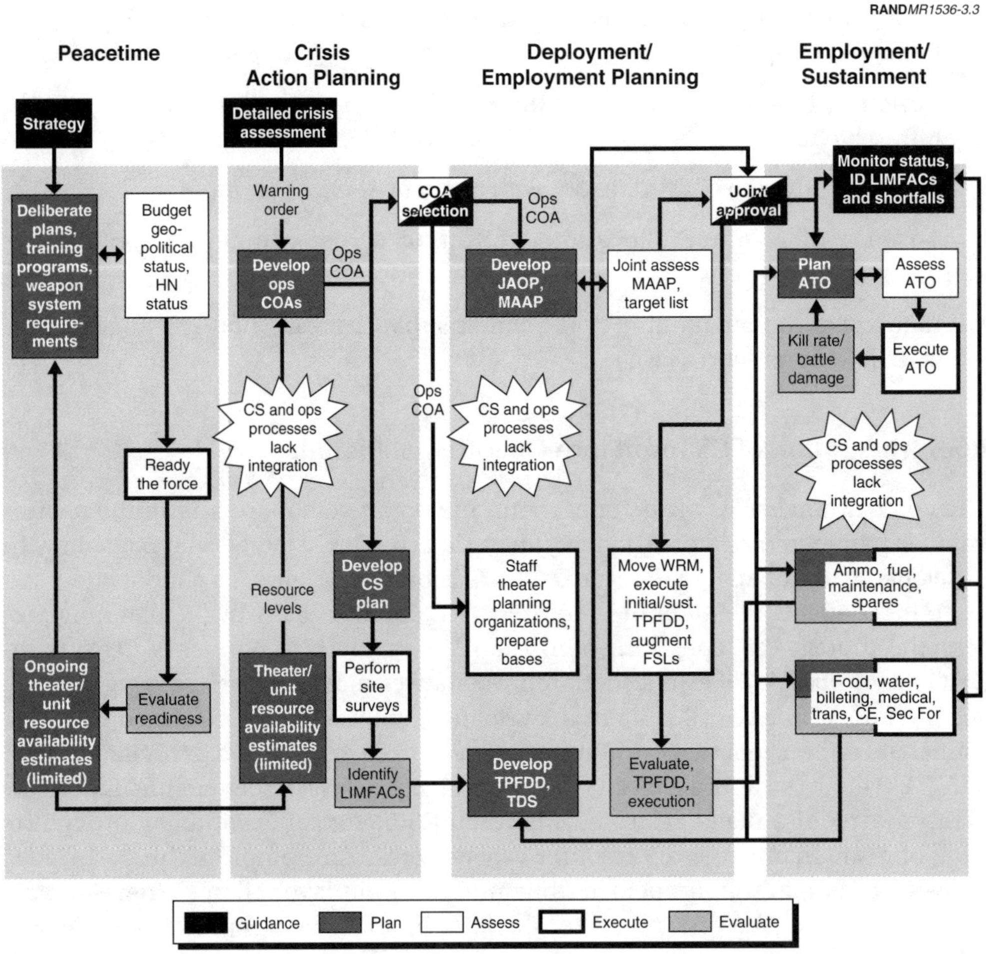

Figure 3.3—CS and Operations Process Integration Shortfalls

understanding of how and when the broad effects of CS considerations enter the planning process.[15] CS personnel are not equipped to communicate essential aspects of CS options in operationally understood metrics. For the most part, the tools to accomplish this translation do not exist. For instance, a delay in setting up resupply could result in sortie degradation, and yet the tools do not exist to translate added resupply time to weapon system availability. As a result, CS information tends to be provided to planners in the form of inventory levels or process performance (e.g., resupply time) rather than base beddown capability, sortie generation capability, or other metrics more relevant to operational planning.[16]

[15]Lt Col Stephen Luxion, Hq CENTAF A-3/A-5, February 8, 2001; Mr. Van Hazel, 7th Air Force operations research analyst, December 10, 2001; Major Parker Northrup, 7th Air Force Air Operations Group, December 10, 2001; Major Steen, PACAF/XPXX, December 17, 2001; Lt Col Levault, 13th Air Force/A-3/5, December 13, 2001.

[16]Lt Col Stephen Luxion, Hq CENTAF A-3/A-5, February 8, 2001; Mr. Van Hazel, 7th Air Force operations research analyst, December 10, 2001; Major Parker Northrup, 7th Air Force Air Operations Group,

At the same time, operators lack logistics or installation support training, and hence tend not to consider the effect that support capabilities have on planned missions. When reliable CS information throughout the operational planning process is not valued, CS aspects of plans are likely to be overlooked, resulting in overly optimistic operational plans that may have to be altered during execution.

An additional hindrance to incorporating CS input into operational planning is a lack of CS assessment capabilities and up-to-date and reliable CS resource information. Assessments may be available for some high-priority situations as a part of the deliberate planning process, but there is no standard, quick procedure for conducting assessments. Even where good logistics or installations support assessment tools exist, they are unique to that command and not readily accepted or interoperable with other MAJCOMs. Thus, when the situation departs from deliberate planning, the system has difficulty making the appropriate assessment quickly and adapting accordingly. Departures from planning can lead to the command "flying blind."

A commonly described shortcoming of crisis action planning was that operators had to plan with incomplete logistics and base support data.[17] As a result, aspects of plans were often made based on outdated information and assumptions, with the CS information typically requested in piece-meal fashion as it became necessary.

Absence of Feedback Loops and the Ability to Reconfigure the CS Infrastructure Dynamically

Combat support and operations activities must be continuously monitored for changes in performance and regulated to keep within planned objectives. Today, asset visibility is limited and in-transit visibility is poor. Thus, it is difficult to estimate current resource levels and future arrival times. Combat support feedback data—resource levels, rates of consumption, critical component removal rates, and critical process performance times such as repair times, munitions build-up times, in-transit times, infrastructure capacity, and site preparation times—may not be recorded routinely. Even when these data are available, they are typically the focus of planning and deployment rather than employment and sustainment. Because operations can change suddenly, these data must be continuously available throughout operations in order to make needed adjustments.

When monitoring reveals a mismatch between desired and actual resource or process performance levels, it may be difficult to find the source, particularly for activities supporting multiple theaters (such as depot repair), or multiple services (such as a TDS or construction priority). Discrepancies between desired and actual levels of support may arise from changes in CS performance or in operations. Assessment must be able to quickly address CS performance problems or changes and estimate

---

December 10, 2001; Major Steen, PACAF/XPXX, December 17, 2001; Lt Col Levault, 13th Air Force/A-3/5, December 13, 2001.

[17]Lt Col Stephen Luxion, Hq CENTAF A-3/A-5, February 8, 2001; Mr. Van Hazel, 7th Air Force operations research analyst, December 10, 2001; Major Parker Northrup, 7th Air Force Air Operations Group, December 10, 2001; Major Steen, PACAF/XPXX, December 17, 2001; Lt Col Levault, 13th Air Force/A-3/5, December 13, 2001.

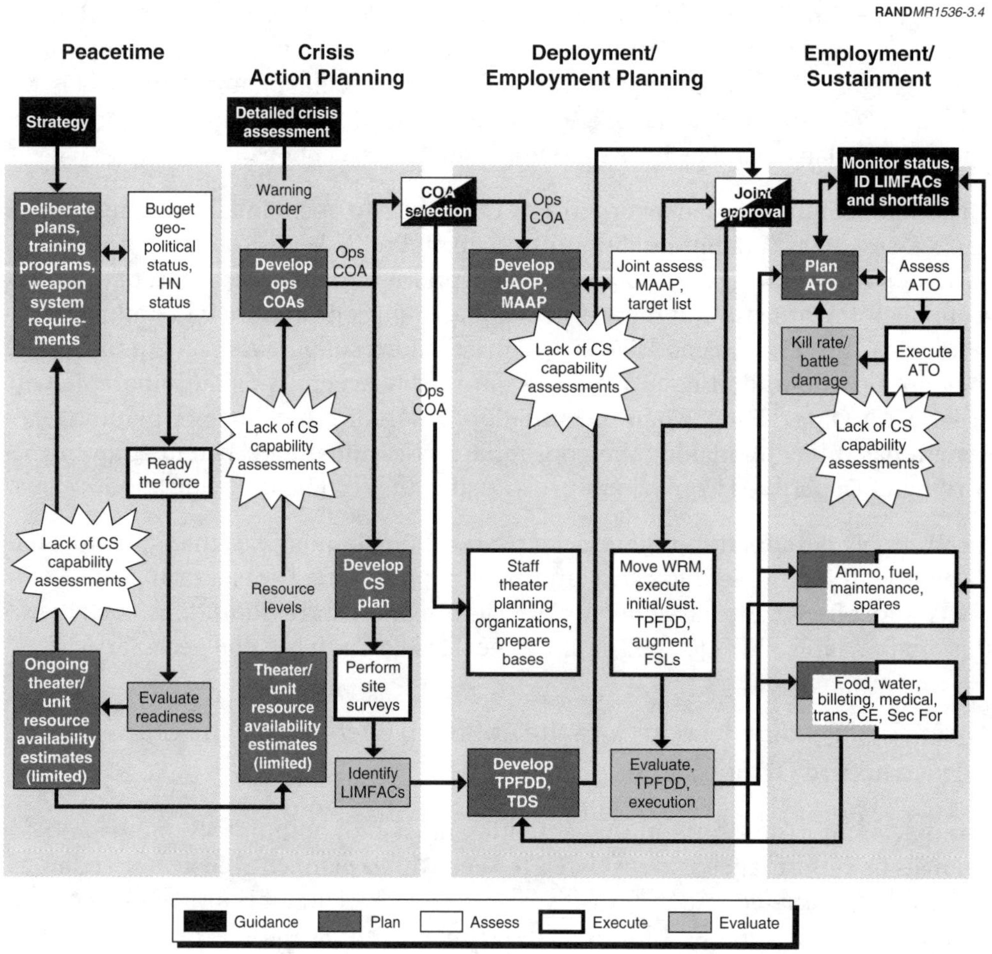

Figure 3.4—Capability Assessment Shortfalls

CS requirements to meet changing operational objectives. Figure 3.4 highlights where the lack of capability assessments affects process execution.

With limited monitoring and performance assessment, it is hard to know when to intervene and adjust CS activities in real time.

Poor Coordination of CS Activities with the Joint/Allied/Coalition Communities

Ultimately, most CS (logistics and installation) activities entail coordination among the services and the joint-service community. Examples include infrastructure repairs, fuels management, the distribution and storage of munitions and housekeeping sets, and transportation. Nowhere is such coordination more important and troublesome than in transportation and distribution management. Inter- and intratheater distribution relies upon the combined efforts of the Air Force, Army, Navy, commercial carriers, and the theater Joint Forces Command, all of which have separate responsibilities and all of which depend on the others for successful operation. Nominally, the Air Force is responsible for providing airlift, the Army for providing surface lift and port management, the Navy for providing sealift, and the combatant commander for theater distribution, often through the appointment of one service component as the executive agent for managing distribution operations.

In principle, the distribution system can operate smoothly if everyone does their job and knows their role; troubles can arise when the relative roles of the different contributors in an operation are not understood, expectations differ on anticipated performance, or priorities differ among the major players. In ONA, the Air Force played a much bigger role than the Army, and there may have been very different expectations for distribution system performance. As part of the AEF, the Air Force relies on deploying rapidly with small amounts of resources deployed with the forces. This structure requires rapid resupply to sustain the forces. However, this requirement was not effectively communicated to the European Command (EUCOM) before the conflict, and EUCOM policies were not set up to support rapid distribution. EUCOM policies required the identification of specific movement requirements before transport assets could be set in motion. The Air Force lean logistics policies had relied on frequency-based distribution rather than on "full transport vehicle" policies that EUCOM was practicing at the beginning of ONA. This difference in expectations and lack of understanding of Air Force needs created distribution problems for the Air Force during ONA. As a result, despite the mature infrastructure available in Europe, the distribution system that supported the Air Force during ONA was slow to start, often relied on ad hoc solutions that bypassed standard procedures, and may not have kept pace with Air Force needs.[18]

Because the AEF relies on rapid distribution logistics and CS depends on rapid and reliable transportation, rapid theater distribution systems should be implemented that take full advantage of cooperation with the Army, Navy, joint-service community, and allied/coalition forces (if applicable). If rapid resupply cannot be established, the Air Force may have to rethink lean policies and deploy with more resources to sustain operations, which would negatively affect deployment and employment timelines.

Just as CS needs and capabilities must be communicated to operations planners, so too must they be communicated, agreed upon, and resourced with other services, the joint-service community, and allied/coalition organizations. In considering intratheater movement to better support the AEF, the Air Force must estimate transportation requirements based on anticipated sortie production goals and understand in what form those requirements should be communicated to the agency responsible for theater distribution. These estimates can be used to help structure demand-based distribution services.

Similarly, CS personnel should clearly define base capabilities to execute beddown plans and be prepared to provide those requirements to coalition/allied forces that may host Air Force units in a contingency. Such communications with al-

---

[18]Feinberg et al., 2002.

lied/coalition forces could accelerate site survey, base development, and beddown planning during the time-critical crisis action planning process. They are essential to laying the foundation for coalition support and participation in execution of beddown and sustainment activities. They are also vital to how command and control of coalition installation support forces is established. For example, in ONA, only when an integrated coalition engineer organization was established with unity of command did beddown activities fully align with theater commander priorities.

Absence of Mechanisms to Facilitate Resource Allocation Arbitration Across Competing Theaters

Resources planned for other regions must often be diverted to support a theater preparing for or engaged in a contingency. However, although the current process can allocate resources among units within a theater, it cannot formally allocate scarce resources across competing theater and JTF demands or support analyses that should accompany requests for scarce resources. As shown in Figure 3.5, the ability

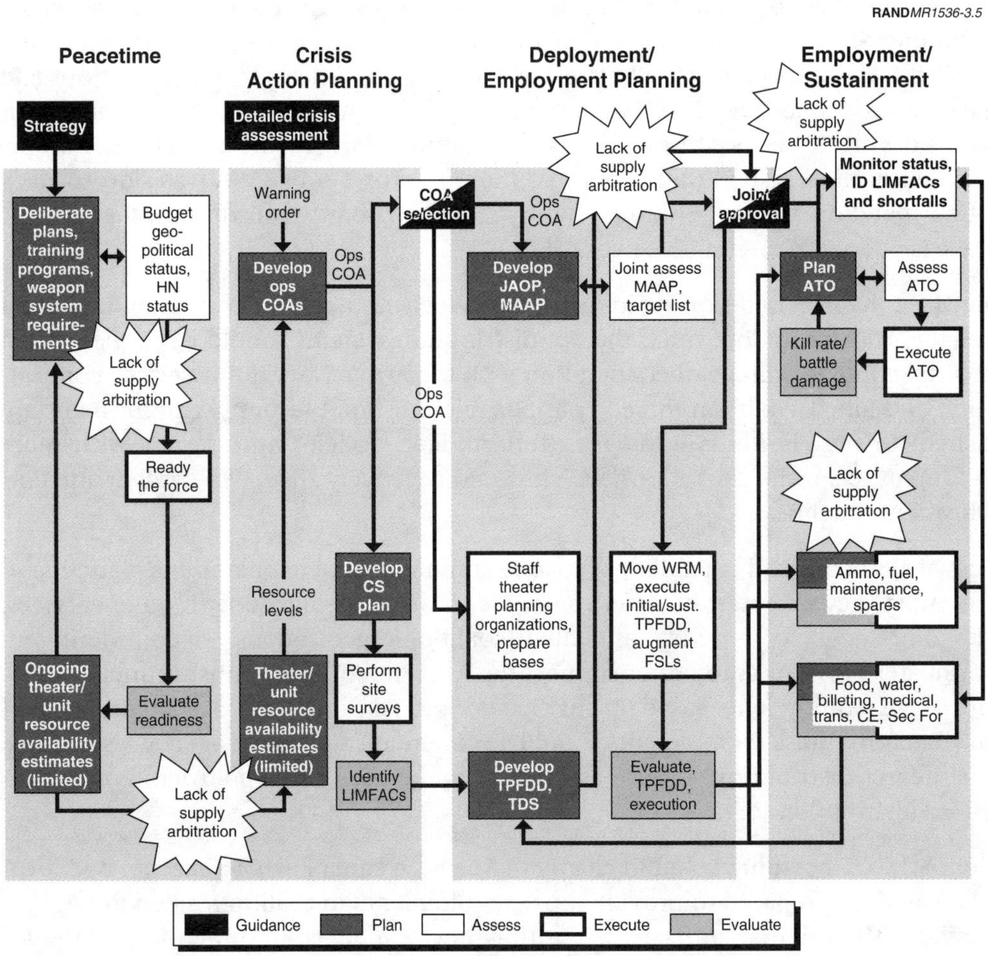

Figure 3.5—Global Supply Allocation Arbitration Shortfalls

to quickly assess the effect on readiness of moving resources from one theater to another is missing across all phases of conflict.

For example, the Ammunition Control Point (ACP) at Hill Air Force Base, Utah, controls the global prepositioning and movement of munitions. However, assessment procedures are not formalized and automated decisiontools may not be in place if they exist at all. How to use them may not be straightforward.[19]

Seventh Air Force can now assess munitions availability in their AOR using standard Air Force munitions computation models.[20] Seventh Air Force A-5 and A-4 organizations are attempting to use this model to show how the reallocation of smart munitions from their AOR to Operation Enduring Freedom (the war in Afghanistan) will affect other war plans. And 5th Air Force engineer planners can now assess and adjust explosive storage capacity in near-real-time to assist in rapid decisionmaking of munitions relocation.[21] This type of assessment must be done before resources are reallocated so that high-level decisionmakers (up to and including the JCS) can see the effect of their allocation decisions before the fact.

Inadequate Understanding That Combat Support Refers Not Only to Logistics But to Installation Support as Well

Attempts to incorporate CS inputs into operational planning not only faced the traditional separation between operations and logistics but also the separation between logistics and installation support. Logistics and their installation support counterparts grow from experience and training in two very different career paths. It is false to assume that in a contingency logisticians or installation support can rapidly become well-versed in each other's diverse activities.[22]

For example, during Operation Desert Shield the 3rd Air Force staff was quickly overwhelmed by requests for detailed logistics information for supporting units deploying to Saudi Arabia at the same time the staff was trying to build up contingency bases for B-52s, tankers, and contingency hospitals in the UK. A split AFFOR role evolved, with 3rd Air Force/Logistics (3AF/LG) having the lead as A-4 for combat support flow to Southwest Asia and 3rd Air Force/Civil Engineering (3AF/CE) having the lead as A-7 for contingency base activation in the UK.

Similarly, during JTF Noble Anvil (NA), 16th Air Force staff initially assumed AFFOR activities in their AOR. However, as the force laydown grew, it flowed to non-16th Air Force locations, forcing the small 16th Air Force staff to rely on HQ USAFE and 3rd Air Force staffs to fully manage CS for NA missions. Eventually, the expanded combat force size led to USAFE/LG assuming the NA AFFOR A-4 logistics responsibilities and the USAFE/CE taking on the NA AFFOR A-7 civil engineer/installations role.

---

[19]Lt Col Carl Puntureri, JCS/J-4 Munitions and NBC Defense Equipment, February 23, 2001.

[20]Mr. Van Hazel, 7th Air Force, Osan Air Base, Korea, December 10, 2001.

[21]Col Brian Fisher, 5th Air Force A-7, Yokota Air Base, Japan, January 15, 2002.

[22]MGen Marcus Andersen, 3rd Air Force/CC, Operation Desert Shield, October 1990.

More recently, during OEF, parallel rapid growth in logistics sustainment and base build-up/relocation motivated U.S. Central Command Air Forces (CENTAF) Forward to reinforce A-4 logistics capability while creating an A-7 installations support function. That action provided CENTAF both senior-level experienced decisionmakers and trained staff to team up on the complicated dual CS challenges of logistics and installation support.

Analysis of the CSC2 processes associated with the above three examples showed duplication of some activities when these CS functions acted independently but synergistic improvement when they teamed up. For example, initial CENTAF AFFOR logistics plans preliminary site surveys did not match up with engineer runway, parking, and infrastructure estimates. But when the information was combined in an automated tool,[23] rapid beddown planning flowed superbly. Another example was A-4 logistics dilemmas with fuel off-load, flow, and storage at a few Southwest Asia basing locations. When CENTAF A-4 integrated technically feasible COAs for solving the urgent fuel dilemma, with inputs from ACC CAT, A-4/7, and CENTAF A-7, a mission solution was quickly identified and executed in half the estimated time.[24]

Thus, CS needs must (1) be managed by staff with adequate depth/experience/rank and (2) integrated with CSC2 processes to focus the results.

---

[23]See footnote 5 in Chapter Four.

[24]BGen Patrick A. Burns, ACC/CE, February 26, 2002.

Chapter Four

CS EXECUTION PLANNING AND CONTROL *TO-BE* CONCEPTS AND OPERATIONAL ARCHITECTURE FOR THE FUTURE

There are ways to mitigate the process disconnects identified in Chapter Three. The *TO-BE* concepts described in this chapter integrate operational and CS planning in a closed-loop environment, providing feedback on performance and resources.[1] Figure 4.1 illustrates these concept elements in a process template that can be applied through all phases of an operation from readiness, planning, deployment, employment, and sustainment to redeployment and reconstitution. The figure centers on integrated operations/CS planning and incorporates activities for continually monitoring and adjusting performance.

Some elements of the process, shaded in medium gray in Figure 4.1, take place in planning for operations and should be accomplished as concurrently as possible. A key element of planning and execution in the process template is the feedback loop that determines how well the system is expected to perform (during planning) or is performing (during execution) and warns of potential system failure. It is this feedback loop that tells the logistics and installations support planners to act when the CS plan and infrastructure should be reconfigured to meet dynamic operational requirements, during both planning and execution. The CS organizations will need to be flexible and adaptive to make changes in execution in a timely manner.

The feedback loop not only drives changes in the CS plan but might call for a shift in the operational plan. For the CS system to provide timely feedback to the operators, it must be tightly coupled with their planning and execution processes and systems and provide options that will result in the same effects yet cost less in CS terms. Feedback might include notification of missions that cannot be performed because of CS limitations.

Figure 4.2 shows how the *TO-BE* concepts can be applied to each phase of a contingency. More detail on the *TO-BE* process can be found in Appendix C. From readiness through redeployment and reconstitution, the core process remains the same, but individual information flows vary and plans and assessments become

[1]Elements of these concepts were described in the Air Force C2 Aerospace Command and Control, Intelligence, Surveillance, and Reconnaissance Center, *USAF Command and Control CONOPs*, Vol. III, *Blue Order of Battle, Global Awareness for Expeditionary Aerospace Forces*, Langley Air Force Base, VA, July 7, 2000, as well as in Ray Pyles and Robert Tripp, *Measuring and Managing Readiness: The Concept and Design of the Combat Support Capability Management System*, RAND, N-1840-AF, 1982.

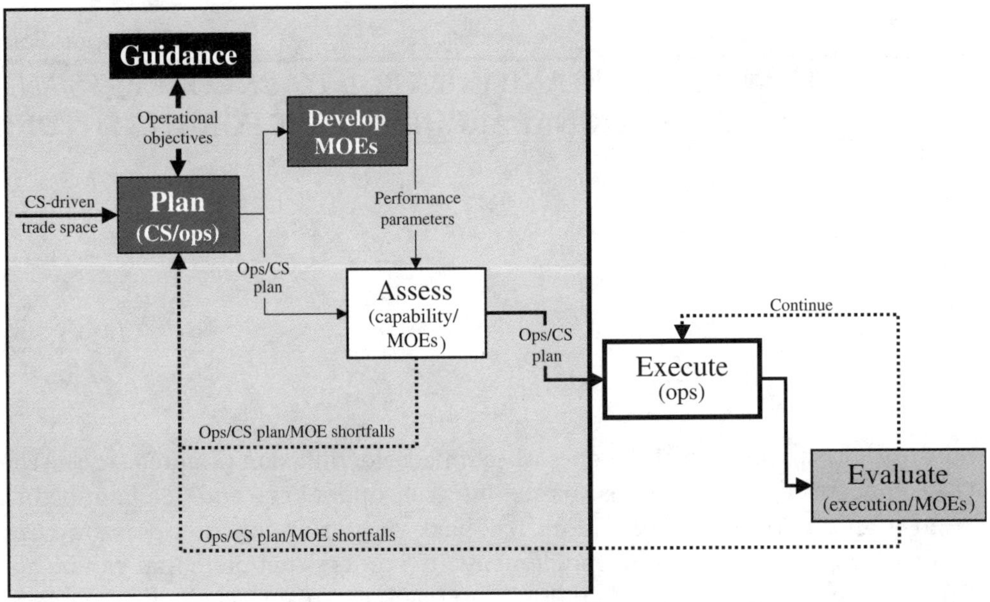

Figure 4.1—CS Execution Planning and Control *TO-BE* Concept

more refined through each phase. For example, in the *TO-BE* environment, theater and unit capability is constantly being assessed, beginning in peacetime. The assessment results are input to the budgeting and planning processes to allocate funds to programs and redistribute other resources to support Air Force plans. The global-level assessment results will contribute to strategic resourcing decisions. As a world situation develops, the CS-driven trade space of operational capabilities feeds into the crisis action planning process and the development of a suitable COA. Based on new information (e.g., refined operational requirements, known threats, better known theater capabilities), the CS plan is refined and the infrastructure configured to support a new COA. As a result of the chosen COA and CS configurations, the trade space is refined to feed into the development of the JAOP and MAAP and eventually the ATO. Assessment capabilities and a feedback loop enable iterative planning. This process continues into employment and sustainment, and can be observed for the other blocks in the planning and execution process.

The detailed *TO-BE* process diagrams are found in Appendix C and associated compact disks. The HTML diagram and supporting database on the disk incorporate the essential process and organizational elements of the *TO-BE* operational architecture. The diagram shows primary CS execution planning and control activities, each of which is depicted in greater detail, including tasks and information flows, in the database.

We will now discuss the *TO-BE* architecture and its application in support of planning and execution.

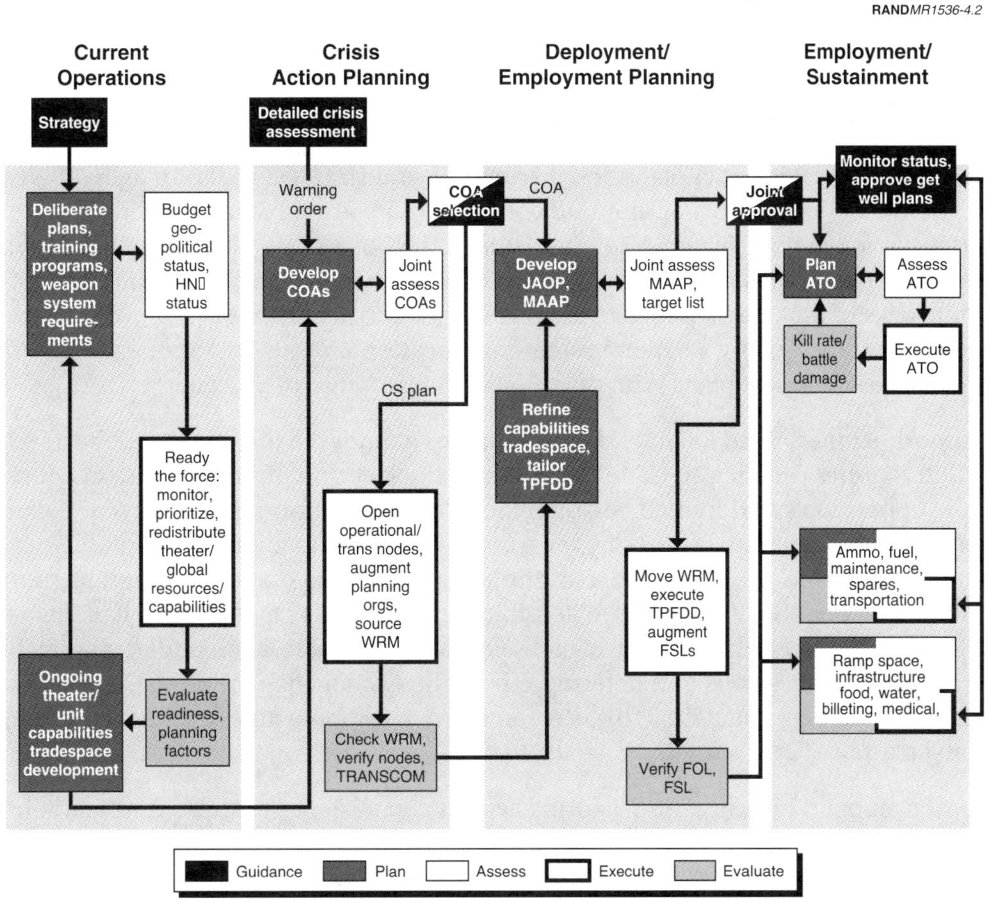

Figure 4.2—Mid-Level Detail of *TO-BE* Process

STRATEGIC PLANNING

The planning activities reflected in Figure 4.2 occur across the spectrum of operations. During day-to-day operations, planning supports programmed flying hours to achieve training objectives and prepare for combat. Planning products are flying schedules and air campaign plans for the operators. For logistics support, they include depot maintenance repair plans, spares allocation plans, and war reserve materiel distribution to support the flying program and air campaign plans. On the installation support side, planning products center around infrastructure operation and maintenance, utility operations, and personnel service activities such as billeting and dining. Exceptions are emergency response activities such as fire, Explosive Ordnance Disposal (EOD), and medical. During wartime or contingency operations, combat execution is prepared in the crisis action planning process, with similar products and plans produced quickly. For both peacetime and wartime planning, we focus on the CS aspect and identify interaction with operators.

The first step in planning is to estimate CS resource (e.g., fuel, munitions, personnel, facilities, equipment, etc.) needs based on the operational requirements, which are typically defined in terms of required sorties by weapon system type. Care must be taken to incorporate uncertainty into the planning process.

Given an uncertain set of operational scenarios and strategic goals, an agile, robust CS system for execution planning and control should be able to meet a wide range of potential outcomes. Figure 4.3 demonstrates the operations planning closed-loop system and identifies the various products of each phase of the process at the operational level of war. Each of those products will identify factors for the CS plan. Sortie rates and durations by weapon system and location are the operations data most critical to the support plan. With those data, CS planning can proceed.

Support planners need to know U.S. capabilities at both the theater and global level, which requires centralized CS information to track each commodity resource level and support tools and trained personnel to aggregate the resource reports and convert them to operational capability measures. With the capability measures, CS personnel can assess the feasibility and implications of each operational and support option, and present a trade space of feasible support options to operational planners. With this trade space, the planners can select a strategy with a full understanding of its support implications and determine how the CS network should be tailored to best fit the chosen scenario. This is essential to developing an effects-based operational plan.

Combat support infrastructure tailoring actions can take many forms. Configuration actions can address the use of CIRFs, development of the distribution network, or the identification of sources of supply (SOSs), to name just a few. The support plan should establish inventory levels for such commodities as spare parts, munitions, and fuel, including safety stocks, at each node of the CS infrastructure, and it should provide protection against uncertainty.

Other elements of the CS plan are the expected performance of the CS infrastructure and the expected consumption of resources based on the planned operational tempo. Planning factors include parking capacity of aircraft ramps, potential fuel consumption versus available fuel storage, critical water and power capacities, expected removal rates for reparables, expected repair times for commodities through the various repair facilities, expected response times at various points within the distribution network, and expected munitions expenditure rates. These planning factors become critical inputs to the decision support tools that provide the "look-ahead" capability that enables combat support to be proactive.[2]

---

[2]Recently, 7th Air Force has developed a modeling capability, using an Air Force model called THUNDER, to determine the effect of critical munitions availability on their war plan. This capability is a positive step for incorporating combat support modeling into OPLANs, as discussed above. In their assessments, 7th Air Force planners will examine how reallocation of smart munitions from the Korean AOR to the air war over Afghanistan will affect movement of the Forward Edge of the Battle Area and additional aircraft and sorties that might be needed to compensate for the lower effectiveness of fewer smart bombs used against targets. They then will explore bedding down the additional aircraft on the Korean Peninsula.

RANDMR1536-4.3

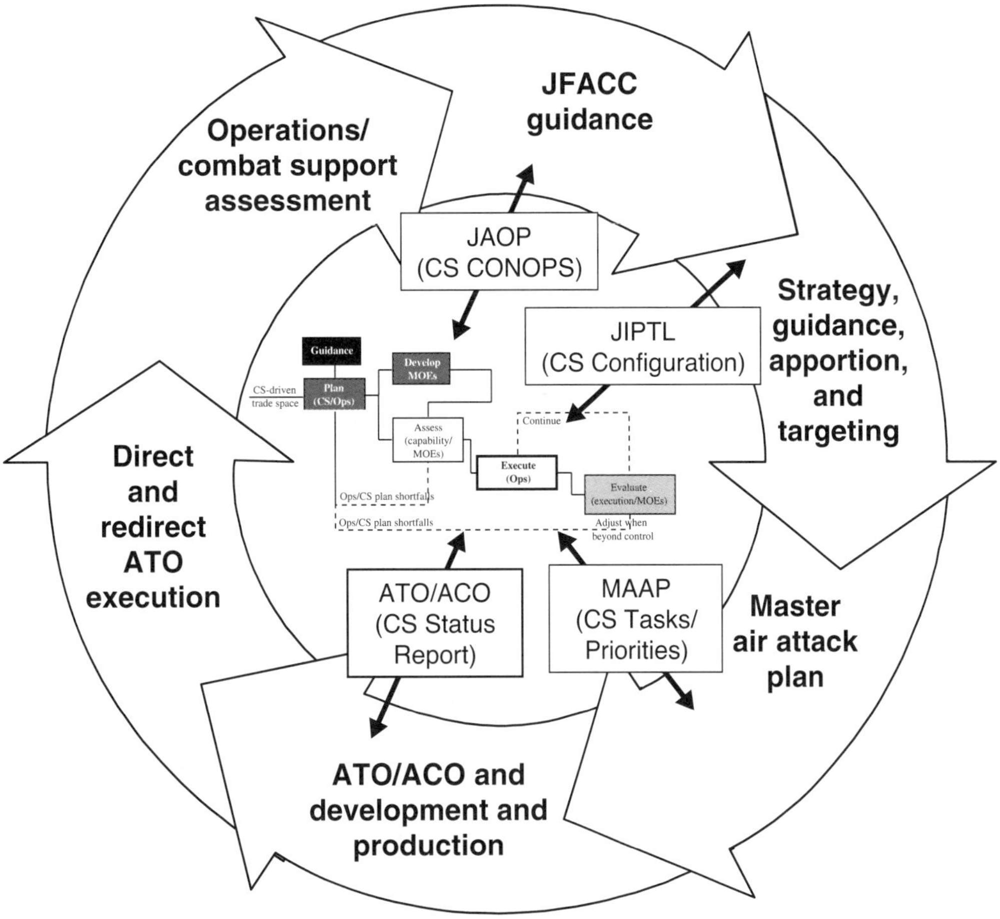

SOURCE: Air Force Instruction 13-1AOC, Vol. III, *Operational Procedures*, Air Operations Center, Washington, D.C., June 1,1999.

Figure 4.3—Integrated Operations/CSC2 Processes

The support network is configured from these plans. Most configuration takes place at the start of the execution phase, although some preparation of the battlefield for contingencies are carried out during strategic planning. Consolidating beddown of like aircraft types, resourcing theater distribution assets, prepositioning war reserve materiel, creating standing centralized maintenance facilities, or establishing command and control nodes are just a few examples of strategic configuration.

In some cases, the Air Force is not responsible for configuring elements of the infrastructure and requires capabilities managed by other services or agencies. The *TO-BE* concept supports Air Force requirements by providing information to the resource executive agent (e.g., the requirement for intratheater transportation to support movement of parts to and from centralized maintenance facilities). Similarly, the need for host nation resources in support of base access and force beddown should be communicated to the authority managing the allocation of those resources. In the *TO-BE* concept, Air Force requirements for the resources will be communicated to the appropriate agency to be considered when configuring those resources.

The next step is to assess the capability of the configured infrastructure. Staff determine whether airfield capacity (ramp, fuel, munitions, power, water), inventories, supply sources, repair facilities, and the distribution network can support operational requirements. Anticipated shortfalls will require retailoring the infrastructure configuration plan.

The plan is assessed against a set of metrics tied to goals such as sortie production capability. If no feasible CS plan can be created within reasonable cost, CS leaders must provide alternatives. The alternatives should provide the same or similar effects as in the original plan. The plan-assess-replan iterations continue until a feasible solution is found.[3]

The final step is to define any further configuration actions as the plan is executed. As mentioned earlier, configuration actions in the support plan are expected to consider the dynamic nature of operational requirements and the resources needed to mitigate the risk of variability in forecasted demands and in the CS processes.

Safety measures to mitigate these risks will be dependent on the scenario and commodity. For instance, increased inventory at strategic locations can serve as safety stock. Additional Line Replaceable Units (LRUs) at the FOL would alleviate the burden on the transportation system if intermediate repair is at a central location. For installation support, increased backup storage for fuel, water, or food can add days to a safety reserve. Increasing capacity at an intermediate repair location would accommodate higher than expected LRU removal rates. Forward positioning of intratheater lift capacity would ensure transportation to move low-demand, high-value parts. If intratheater transportation is the critical resource and might be a potential CS bottleneck, forward resource positioning at the FOL can alleviate demands on the distribution system. (That would be unnecessary, however, if there is an abundance of intratheater transportation and low demand from other services.) Each of these safety measures should be considered and an appropriate set worked into plans. CS planners need tools to assess these types of safety measures.

PLAN EXECUTION AND PROCESS MONITORING AND CONTROL

Plan execution includes peacetime activities (flying training missions or moving materiel within CONUS) and wartime activities (carrying out the air campaign plan, and deploying CS materiel to the theater). These activities are dictated by operational objectives, resource requirements, and configuration actions developed in the planning phase.

---

[3]The 7th Air Force munitions example in the previous footnote provides an insight into the type of assessments required here. We note that this type of assessment is needed for contingencies and unplanned wars and is not limited to canonical planning scenarios.

Once combat operations commence, the logistics and installations support infrastructure must be regulated to ensure continued support for dynamic operations. The system must monitor actual CS performance against that planned. The performance parameters and resource buffers established during execution planning will provide advance warning of potential system failure. When CS performance diverges from the desired level, the system must be able to detect the change, modify the original plan, develop a get-well plan, and reassess the modified plan's feasibility. Plan feasibility is assessed continuously. Safety measures, inventories, and high-level metrics are key elements in CS monitoring and control.

High-level metrics are shown in Table 4.1. Metrics at the command echelon can warn of a pending inability to meet operational requirements. Metrics at the mid- and lower level are tied to the higher-level metrics but enable adaptive planning at the lower echelons and should provide earlier warning of potential problems. The linkage of metrics across command echelons ensures that performance at the lower echelon is tied to higher-level operational requirements. At the same time, it allows lower echelons to monitor subsystem metrics and make corrections that do not sub-optimize at the lower level at the expense of higher-level requirements.

Within the metrics hierarchy are a few key decision measures: operational cost of CS performance shortfalls and CS cost of operational objectives. The CSC2 system must support metrics for the operational/combat support tradeoff. The look-ahead analysis must address the long-range impact of near-term decisions from both an operational and a CS perspective.

As the system monitors performance, it must indicate when key operational measures are out of control and then facilitate get-well planning. When early warning of an impending failure to support operational requirements is received, the system should be able to drill down to the element or infrastructure component that

Table 4.1

Hierarchy of CS-Related Operations Metrics

Command Level	CS Performance Measure	Decision Measure (Ops/CS tradeoff)	Ops Performance Measure
High Level (JFC/JFACC/AFFOR)	Fleetwide MTW readiness Theater infrastructure preparedness	Operational cost of CS shortfalls Combat support cost of achieving operational objectives	Strategies accomplished: Centers of gravity Selected force availability
Mid Level (AOC Dir/A3/A4)	Weapon system availability Critical asset inventories		Tasks accomplished
Low Level (Wing/Depot/AFMC/Battlestaff)	Order & ship times Expected back order Depot repair cycle time Base repair cycle time Inventory levels		Targets killed per sortie Missions launched Other

NOTE: JFC = Joint Forces Command; AFMC = Air Force Materiel Command.

is contributing to the general failure. While the system is being monitored at the higher level against key operational measures, the lower levels are monitoring the performance of component processes against the planning parameters and thresholds established during execution planning. Because of the breadth of the resources and processes at the lower levels, system performance must be monitored with a few key high-level indicators, supplemented by a detailed metrics hierarchy at all echelons. After drilling down to the root cause of a problem, each echelon can manage effectively.

Finally, the *TO-BE* concept permits allocation of critical resources among competing demands. Resource arbitrations are carried out by the allocating authority. The request and associated cost/benefit trade space will be presented in operational terms related to the strategies and objectives initially communicated by the guidance authority.

AN EXAMPLE OF CS EXECUTION PLANNING AND CONTROL IN A SMALL-SCALE CONFLICT SCENARIO

We use an example of a small-scale conflict to discuss the *TO-BE* concept in both planning and execution.

In this scenario, Iraq attempts a lightning strike to seize key objectives in Kuwait and Saudi Arabia. The United States will attempt to defend Saudi Arabia and Kuwait using the predeployed Southern Watch force, augmented with an additional AEF on Day 3. We will apply the *TO-BE* concept in selecting, deploying, and sustaining the AEF.

Planners are considering two force mixes that are both capable of providing the desired results. The first is a force composed of four fighter squadrons (each with 24 aircraft) and three B-1 bombers deployed to bases in Saudi Arabia. The other force consists of a 24-aircraft squadron of fighters and 24 B-1 bombers operating from bases outside the theater. Given that both forces can achieve similar, acceptable effects (the first force can likely halt the attack at Kuwait City, the second force can likely halt the attack at the Saudi border), other factors such as deployment response time, beddown infrastructure requirements, and munitions and fuel requirements might enter into the force selection.

With weapon system selections derived from operational strategies and objectives, CS planners will review a list of beddown sites. Up-to-date knowledge of FOL and host nation capabilities will help determine which sites have sufficient runway length, aircraft parking space, tarmac load-bearing capacity, fuel and munitions storage capacity, and munitions and fuel inventory (if applicable). From the initial list of potential beddown sites that can support a variety of aircraft, options narrow; however, additional information is needed on sortie rates and munitions types to evaluate the sites' ability to support mixes of force types and weapons systems. This iterative exchange of requirements and capabilities information between operations and CS planners is characteristic of the future environment, which will demand a system able to quickly assess force-mix beddown requirements against available intra- and intertheater beddown capabilities.[4]

As planners analyze the force mix, they may arrive at different combinations and numbers of weapon systems that can meet the operational objectives within politically acceptable strategies. In this example, force employment analysis reveals two force packages that can provide the needed firepower: one requires three beddown sites and the other requires two. In continuing the iterative force beddown planning analysis, a CS execution planning and control system must be able to examine site capabilities and their ability to support the operational planners' force mixes. Critical to the analysis is the ability to drill down into specific capabilities and commodity inventory levels (e.g., fuel and ammunition, fuel distribution, ammunition build-up and distribution, etc.) at each site and to analyze the various force-mix demands against the capabilities over a specified period of operations. This additional, more detailed analysis again requires more information from operations planners with respect to the intended sortie rate and sortie duration for the different force mixes, as well as some general idea of the types and quantities of munitions to be used over the planning horizon.

Given the additional operational planning factors, the system must be able to again quickly compute estimates of resource requirements for force employment options and assess their feasibility against available resources, both in place at the beddown locations and potentially available as part of the force deployment and sustainment plan. A key output is the translation of CS resources and capabilities in terms reflective of operational goals (e.g., sortie rates). As reflected in Figure 4.4, it is important to view the beddown feasibility assessment from both a resource perspective (e.g., gallons of fuel and number of munitions required and available) over time and an operational perspective (e.g., sortie production required and capable) over time. This analysis will contribute to the final selection of the force mix that will be deployed to support operational objectives.[5]

Parallel beddown analysis considerations for installation planners extend beyond sortie production and sustainment to include base support infrastructure/resources to support the population growth associated with force deployment. Base infrastructure analysis must consider the requirement for and availability of resources such as lodging, meals, power, water, and sanitation. As with the sortie production and sustainment analysis, the base infrastructure analysis should be conducted over the time horizon specified by the operators and include both in-place resources and those that can be made available through the deployment and sustainment plan.

---

[4]ACC and PACAF developed a remote beddown assessment capability called GeoReach that was unveiled at CORONA, fall 2000. It relies on ISR assets and a five-step process to locate, image, assess, map, and enable a 70 percent to 80 percent planning solution for potential beddown sites. The GeoReach capability was used in the early days of Enduring Freedom to assess and prioritize potential beddown locations in two to three days versus the two to three weeks it took in earlier contingencies.

[5]The operational assessment should reflect the integration of CS resource assessments. In our example, shortfalls in sortie production are projected on Days 4–5 and 10–15 because of insufficient munitions inventory, and on Days 6 and 12 as a result of insufficient POL (petroleum, oil, and lubricants) inventory.

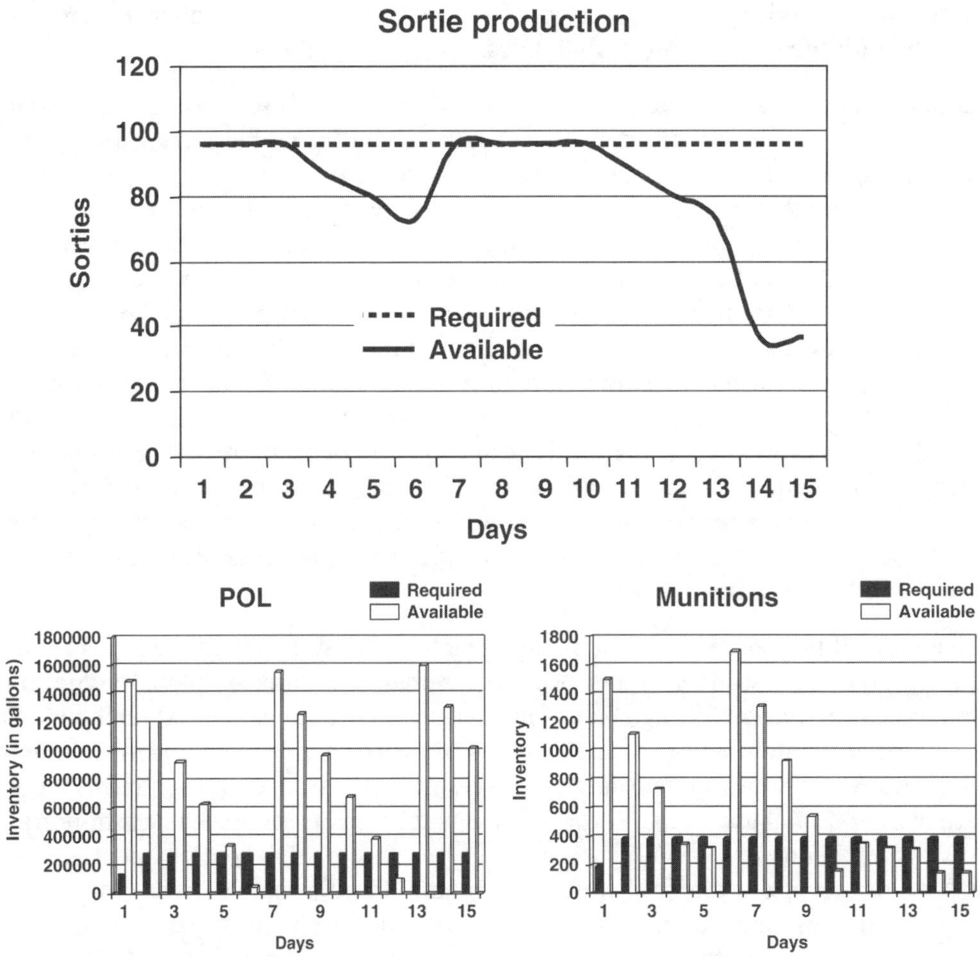

Figure 4.4—Sortie Production and Resource Views

Force and beddown analysis will result in (1) a feasible plan for the placement of forces at FOLs, employment of those forces, and sustainment of both the weapon systems and associated support personnel or (2) an infeasible plan. As shown in Figure 4.5, the sustainment plan should address where and how maintenance operations will be conducted, how inter- and intratheater transportation will be provided, as well as infrastructure configuration and expansion actions to ensure timely support of operations.

As the plan is executed, the focus shifts to resource and process monitoring and control. Our example continues by looking at the role of CS during execution.

In this example, a set of inventory levels has been established for LRUs that cross a common test station at a CIRF. There are only two LRUs that use a common test station, LRU #1 and LRU #2. The LRU levels at the CIRF are dependent on demand rate, resupply time from the intermediate repair facility, repair cycle time at the

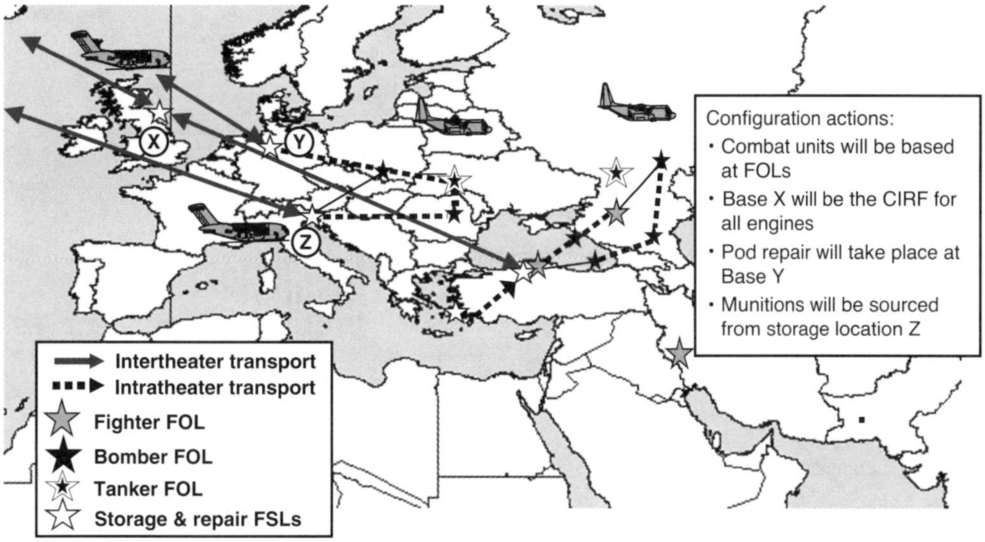

Figure 4.5—Configuration Actions Resulting from CS Planning Analysis

repair facility, and so forth.[6] In this case, the inventory levels are based on best estimates of the expected 30-day flying-hour profile of the weapon system that uses the LRUs. Thus, the inventory levels of these LRUs can be monitored, as can the input parameters used to establish levels, to determine if the flying-hour program is likely to be supportable. Assume an inventory level has been established for LRU #1 that would support "normal" variations in removal rates at the flight line, normal or expected resupply times, and repair cycle times, among other parameters. In the example in Figure 4.6, the level at the CIRF for LRU #1 is projected to drop below its threshold level based on scenario planning, as shown on the upper right side of the chart. Given this projection, the FOL inventory for LRU #1, shown on the lower right side, will go to zero on Day 15, which will result in reduced sorties shown on the left side of Figure 4.6. In this way, the CS execution planning and control system indicates how CS lower-level metrics can be monitored and threshold shortfalls related to combat performance in advance of effects of the shortfalls being realized.

The scenario reflected in Figure 4.6 shows that at Day 15, the fully mission capable (FMC) aircraft percentage is projected to be below acceptable operational thresholds, resulting in required sorties not being flown. Continuing our example, Figure 4.7 illustrates drill-down of LRU #1 performance from the CIRF perspective. This illustration shows how a C2 system component at the CIRF could track key parameters that drive the LRU inventory level, including shop replaceable unit (SRU) levels, repair cycle time, and removal rates at the FOL. The removal rate for LRU #1 is underestimated, given the sortie production surge on Days 11–14, and as a result the FOL

---

[6]See Richard Hillestad, *Dyna-METRIC: Dynamic Multi-Echelon Technique for Recoverable Item Control*, RAND, R-2785-AF, 1982; and Raymond Pyles, *The Dyna-METRIC Readiness Assessment Model: Motivation, Capabilities, and Use*, RAND, R-2886-AF, 1984.

RANDMR1536-4.6

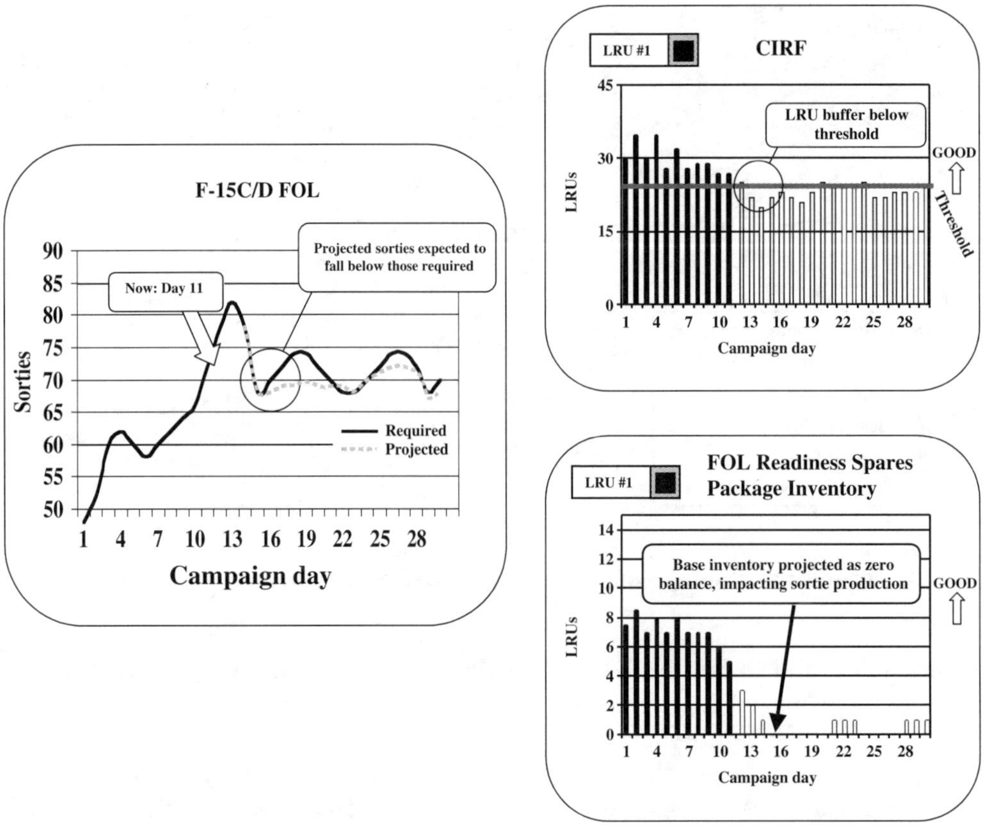

Figure 4.6—Sortie Production Capability and LRU Inventory Level

inventory for LRU #1 is projected to go to zero. The quantity of LRUs #1 at the CIRF will fall below their threshold buffer level, as shown on the left side of the figure (replicated from Figure 4.6).

There are many possible causes for the LRU #1 removal rate to exceed its buffer threshold level. The initial failure rate estimation for this LRU could have been flawed, external factors such as operating in a harsh environment may not have been fully considered, or the failure rate may vary from that planned with respect to the total number of sorties flown. In this case, the removal rate increase was a result of higher-than-expected sortie requirements.

The cause for the decrease in LRUs #1 below the buffer threshold can now be expressed in terms of the effect on projected performance for both operational and logistics decisionmakers. Get-well plans can be developed and implemented. In this case, the level for LRU #1 at the FOL could be adjusted upward and more stock moved from the CIRF to the FOL to protect against the unexpected rise in the removal rate.

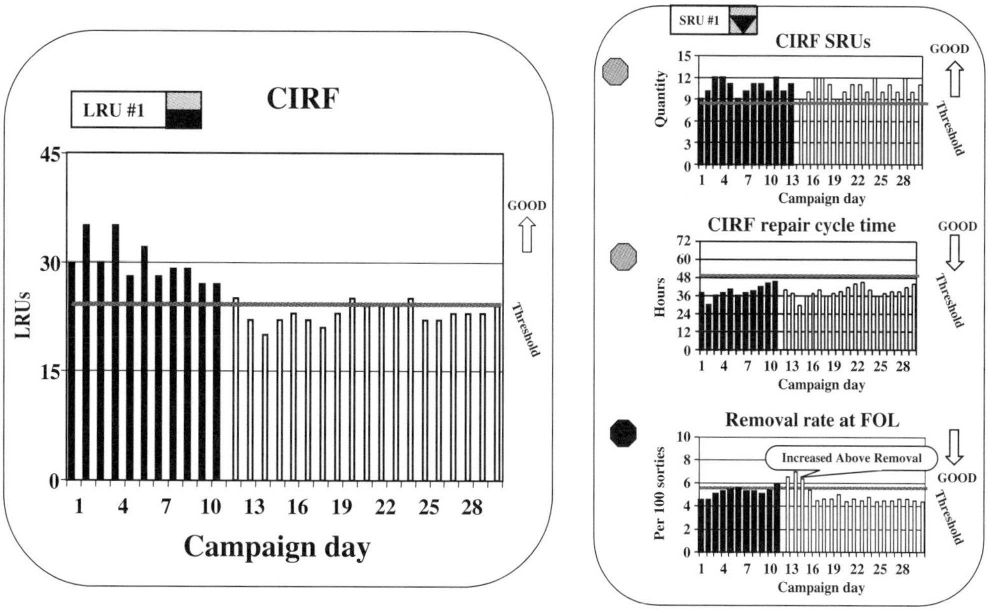

Figure 4.7—CIRF Capacity Drill-Down

By establishing thresholds on performance and appropriate safety-level stocks to respond to changes in operational capabilities, the ability to drill down and pinpoint the potential and actual causes of sortie-generation problems, and a feedback loop to communicate with operations and CS planning organizations, the C2 for combat support system can detect and resolve CS disconnects and enable continued operations.

Although this example deals with a single repair location problem and has a short-term, easily implemented solution, the system and decisionmaking methodology can be used to monitor all aspects of a campaign. For any operational problems discovered during a status evaluation, the execution planning and control system should have tools and personnel to identify the CS shortfalls responsible. These shortfalls can include any aspects of combat support, from repair prioritization to theater distribution or infrastructure configuration. For any problem or proposed solution, the system should be able to project its effects on overall operational effectiveness. Planners must be equipped to choose a solution that addresses the problem at hand, while considering the impact on future risk and capabilities.

Chapter Five

SHORTCOMINGS AND PROPOSED CHANGES

The *TO-BE* concept discussed in Chapter Four has an execution planning and control process designed around the needs of the AEF: operationally relevant, rapid, and responsive. As shown in Chapter Three, several aspects of the Air Force's current process prevent the *AS-IS* operational architecture from providing the most effective support possible to the AEF. Process improvements are thus core to the evolution of an execution planning and control system capable of ensuring support to the AEF. Several "enabling mechanisms," including doctrine and policy, organizational responsibilities, information systems, and training and education, must be modified.

In this chapter, we identify shortfalls in the current system. Some shortfalls became apparent from our *AS-IS/TO-BE* comparative analysis. Some were exposed in previous Air Force operations, and some have been evident in the difficulties of day-to-day activities.

For each shortcoming, we propose modifications targeted at implementing the *TO-BE* concept and demonstrate how these proposed changes will lead to an improved decisionmaking structure.

DOCTRINE AND POLICY

Our analysis revealed several shortfalls in Air Force CS doctrine and policy. The shortfalls and proposed solutions are summarized in Table 5.1.

Although doctrine clearly defines the extent of operational, tactical, and administrative control that operational organizations have over each other, no such definition exists for CS.[1] Because doctrine is minimal for CS, operational planning may not reflect CS realities,[2] delaying plan development, slowing the response to changing plans, and increasing the risk of running out of critical resources later in the campaign.

---

[1]While AFDD-2 emphasizes operational C2, and Air Force Instruction (AFI) 13-1AOC, Vol. III, defines specific tasks associated with the AOC, parallel documentation does not exist for CSC2 and CS organizations.

[2]Lt Col Stephen Luxion, HQ CENTAF A-3/A-5, February 8, 2001; HQ CENTAF LGM, February 7, 2001; Mr. Van Hazel, 7th Air Force operations research analyst, December 10, 2001; Major Parker Northrup, 7th Air Force Air Operations Group, December 10, 2001, Major Steen, Pacific Air Forces (PACAF)/XPXX, December 17, 2001; Lt Col Levault, 13th Air Force/A-3/A-5, December 10, 2001.

Table 5.1

Doctrine and Policy Shortfalls and Proposed Solutions

Doctrine/Policy Shortfall	Proposed Solution
Objectives/functions not well defined in doctrine	Rewrite AFDD-2, AFDD-2-4, AFDD-2-8 to include basic objectives/functions of CS execution and planning and control and organizational alignment
Lack of AF-wide emphasis on C2 for CS	Increase emphasis on CSC2 role
CS organizational responsibilities for C2 not well defined in doctrine or policy	Develop and write policy for CS execution planning and control
Necessary C2 information flows not well documented for CS	Develop and write policy for CS execution planning and control

Table 5.1 shows that current Air Force doctrine and policy place little emphasis on CS input to operational planning and execution. Combat support feasibility assessments play a significant role in the *TO-BE* architecture, whereas AFDD-2 describes the joint-service operations planning and tasking cycle phases as "Plan, Execute, Assess."[3] There is no feasibility check between planning and execution, and thus the process does not consider Air Force status until after execution has begun.[4] If plans are not supportable, corrective actions become disruptive to combat execution as well as to future plans.

Because the CS execution planning and control concept is not well defined in doctrine, the objectives and functions of C2 for CS and assignment of responsibilities to organizations are not well defined in policy.

Air Force Instruction 13-1AOC, *Operational Procedures—Aerospace Operations Center*, provides guidance for the operation of the AOC and clearly denotes the functions involved in operations C2. It describes the purpose and primary responsibilities of the AOC, detailing the tasks necessary to accomplish them. It shows the command relationships between each division in the AOC, the information each division requires and generates, and the tools each uses to do its jobs. Similar guidance for CSC2 is largely contained in concepts of operations (CONOPs), which lack the directive authority of a doctrine or instruction document. MAJCOM and theaters develop operating instructions and CONOPs, but the documents often differ from one command to the next in approach and process.

AFDD-2-8 specifies four functions of a C2 system: planning, directing, coordination, and controlling, with little detail on the tasks necessary to accomplish these functions or which CS organizations will perform them. There is thus confusion regard-

[3]AFDD-2, p. 74.

[4]Hq Air Force Materiel Command (AFMC), February 21, 2001; Joint Staff J-4, Pentagon, February 23, 2001.

ing the responsibilities of CS organizations[5] and communication networks that are ad hoc.[6]

These problems can be eliminated with a series of changes to Air Force doctrine and policy. First, AFDD-2-8 should include CS details beyond the four basic functions of any command and control system. The following are proposed inputs to AFDD-2-8 describing planning, directing, coordinating, and controlling.

Planning. With the AEF's short timelines and pipelines, it is critical to be able to add CS information to initial planning, giving planners flexibility and confidence.

CS execution planning functions include monitoring theater and global CS resource levels and process performance, estimating resource needs for a dynamic and changing campaign, and assessing plan feasibility. Because capabilities and requirements are constantly changing, these activities must be performed continuously so accurate data are available for COA and operational planning.

Planning also includes assessment and ongoing monitoring of CS infrastructure (FOLs, FSLs, CONUS support locations, TDS, C2 nodes) configurations that support the operations plan. Benefits and drawbacks of various support options (use of FSLs, sources of supply, transportation providers, modes and nodes, host nation support) must be weighed in the context of timelines, operational capability, support risk, and cost. Having complete, up-to-date information on FOL capacities and operational capabilities [e.g., number and type of aircraft mission-design-series (MDS) and munitions] and their support (e.g., on-base repair capacity, fuel availability) allows more CS information to play in early planning stages (such as COA development).

CS execution planning should result in the production of a logistically feasible operations plan, a CS plan that dictates infrastructure configuration, a C2 organization structure, a TDS, and CS resource and process control metrics.

Directing. CS directing activities include configuring and tailoring the CS network and establishing process performance parameters and resource thresholds.[7] Planning output drives infrastructure configuration direction—there must be an ongoing awareness of CS infrastructure and transportation capabilities to feed into operational planning and execution. For example, the speed and precision with which beddown sites can be assessed and prepared (configured) improve with the amount of information available beforehand. Awareness of the precise configuration for various options, in turn, gives planners more speed and flexibility in employment of forces in the face of changing objectives or constraints. The ability to reconfigure the

---

[5]Col Huck Robinson, HQ ACC Battle Staff, February 6, 2001.

[6]Col Huck Robinson, HQ ACC Battle Staff, February 6, 2001; HQ AFMC, May 2, 2001.

[7]Heuristically determined thresholds can be established while more sophisticated expert rules or algorithms are being developed. For instance, Brigadier General Hennessey (AMC/LG) uses zero-balance stock positions coupled with forces supporting an engaged combatant commander as a rule to determine when lateral actions should be taken to resupply a unit "at war." Using this rule, he authorizes the AMC/Regional Supply Squadron (RSS) to reallocate stocks from units with stock to those with zero balances. The idea is to prevent mission degradation by focusing attention on the items that will cause the next mission degradation.

support infrastructure quickly enables operational changes, be they the result of anticipated or unanticipated changes in a scenario. Timely, accurate information and an agile CS system able to execute network configuration decisions would thus allow leaders to respond more quickly or simply to make more informed decisions.

Along similar lines, identifying and using appropriate sources (e.g., ships, supply depots, or host nation contractors) for different commodities (e.g., ammunition, fuel, or spares) and required services (construction, billeting, feeding) allow maximum employment of available Air Force and joint-service resources and the opportunity to balance intra- and intertheater requirements to support all AORs. As operational objectives change, requiring different logistics or installation support, the source can be changed. Also, as operational locations change, the source, as part of the overall CS network, can change to meet the demands more quickly.

Coordinating. Coordination ensures a common operating picture for CS personnel. It includes beddown site status, weapon system availability, sortie production capabilities, and the like. Coordination activities should be geared to providing information to higher headquarters, not necessarily to seek a decision but to create an advance awareness of issues should a higher-headquarters decision be needed at a later date. CS coordination tasks will affect theater distribution, force closure, supply deployment, and allocation of support forces. Each of these activities requires information gathered from a variety of processes and organizations and consolidated into a single decisionmaking framework that delivers decision-quality data to planners and commanders.

For example, to coordinate TDS movements, CS personnel must monitor all parts of the theater, as well as the activities of TRANSCOM, other U.S. military services, coalition partners, and host nations. Similarly, base-level planning is usually dependent on supplies provided by intratheater distribution. To develop supportable plans, operational and support planners must understand what the TDS will provide at any given time. Policy should specify the information to be collected and dictate how it should be gathered and disseminated to organizations for decisionmaking or to maintain situational awareness.

Controlling. During day-to-day and contingency operations, CS control tracks CS activities, resource inventories, and process performance worldwide, assessing root causes when performance deteriorates, deviates from what is expected, or otherwise falls out of control. Control modifies the CS infrastructure to return CS performance to the desired state. CS control should evaluate the feasibility of proposed modifications before they are implemented and then direct the appropriate organizations to implement the changes.

While doctrine must define and establish CS execution planning and control functions and objectives as described above, it should also prescribe which organizations perform these functions. AFDD-2 gives the organizational structure of the AFFOR and AOC, and AFDD-2-4 briefly describes the roles and deliverables of CS functions. Doctrine should further delineate the roles and responsibilities of directorates within the AFFOR, divisions of the AOC, and other support organizations (see the next section). It should include the reporting hierarchy and the communication network between groups. Once the "what" and "who" are delineated in doctrine, AFIs (instructions) should detail "how" the function will be executed by describing tasks performed by each organization, the information that each group should consider in its decisionmaking, and how frequently this information is updated.

As shown earlier, our CS *TO-BE* decisionmaking process incorporates C2 functions and objectives. Figure 3.1 differs in appearance only slightly from Figure 4.1, but the latter reflects a significant shift in Air Force priorities and resulting differences in planning and execution. Elevating the importance of CS execution planning and control in Air Force doctrine can engender enforceable rules for each organization, document information to be shared, and enable a much-improved planning process.

ORGANIZATION

In addition to doctrine and policy shortfalls, our analysis revealed several deficiencies in the Air Force's CS warfighting organizational structure. As summarized in Table 5.2, these deficiencies are rooted in the largely ad hoc approach that the Air Force uses to move from varying levels of contingency support in one AOR to differing levels of contingency support in another.

As we noted above, doctrine and policy do not clearly define and delineate the C2 roles and responsibilities of CS organizations, only that organizations shift to wartime roles and augment staff to manage the increased responsibility. The Air Force has supported one contingency after another for the past ten years—what changes is the AOR and the level of support.[8]

Current doctrine calls for the NAF to be the supported warfighting organization and be supplemented. In the last three conflicts, however, the Air Force has instead augmented the MAJCOM to provide AFFOR Rear responsibilities.[9] The augmentation staff was designated at the last minute, creating several challenges. First, the augmenters lacked familiarity with theater-specific plans, limiting their effectiveness in carrying out responsibilities that rely on in-depth knowledge of threat, host nation, and theater issues. Second, staff augmenters were not always familiar with command-unique policies and procedures (many of which are undocumented and have evolved from personal relationships between staff members and intratheater agencies), and they lacked training on locally developed decision support tools that are prevalent in the absence of standardized information systems. Finally, augmenters may lack experience with the core staff they are joining and hence may not contact the most knowledgeable person when seeking help. The impact of these types of uninformed or ad hoc communication networks can be significant. Without

---

[8]We acknowledge BGen Art Morrill (PACAF/LG) for this point.

[9]Feinberg et al. (2002) document that the USAFE/LG staff served as AFFOR A-4 for AWOS. MGen Donald Wetekam (ACC/LG) in an interview on December 21, 2001, indicated that he served as AFFOR A-4 Rear for Operation Enduring Freedom, while BGen Pat Buras (ACC/CE) served as the AFFOR/A-7 Rear.

Table 5.2

Organizational Shortfalls and Proposed Solutions

Organizational Shortfall	Proposed Solution
Transitional roles and responsibilities are unclear	Establish standing CS organizations with clear C2 responsibilities
Peacetime organizations have difficulty shifting to support one AOR from another	Develop procedures for centralized management of CS support resources and capabilities
Roles of joint-service/combatant commander are unclear	
Warfighting organizations are minimally staffed and rely on poorly trained augmentees	
Commodities are managed by different organizations	

knowledge of established points of contact, augmenters may attempt to solve problems on their own, resulting in delays or errors. They may seek assistance from personnel outside the area of expertise who may not have the most current or accurate information. Command-level functions operating with outdated information then may become reactive, preventing staff from the monitoring, assessing, and reconfiguring that can better position CS resources and reduce uncertainty. Informal communication is also problematic in the context of sourcing materiel for a deployment or operation. When support materiel is not available from the group initially tasked, the planners contact anyone they can find for the needed supplies. By the time the operation ends, it is almost impossible to remember who has provided what or to compensate the supplies' original owners. This complicates redeployment plans and often results in the Air Force leaving supplies in the theater.

Another problem with the lack of standardized responsibilities is that organizations with the same name may play completely different roles, depending on the theater. For example, Air Combat Command (ACC), U.S. Air Forces Europe (USAFE), and Pacific Air Forces (PACAF) Contingency Response Groups (CRGs) and RSSs have different responsibilities thus different standards and measures for their performance.[10] It is therefore impossible to plan a common set of CRG tasks or use a common set of metrics to measure the performance of all RSSs. Similarly, it is impossible to develop a single training curriculum for all CRG or RSS operators, and personnel movement and employment between theaters has become increasingly complicated.

Operations Noble Anvil and Enduring Freedom highlighted the challenges of operating without an established CS organizational structure for C2.[11] During ONA, the CS warfighting responsibilities shifted from the 16th Air Force staff to the USAFE staff.

[10]Interviews at ACC RSS, February 6, 2001; PACAF RSS, March 8, 2001; and U.S. Air Forces in Europe (USAFE) RSS, April 4, 2001; ACC CRG, February 27, 2002.

[11]See Feinberg et al. (2002). MGen Donald Wetekam in an interview on December 21, 2001, indicated that he served as AFFOR/A-4 Rear for Operation Enduring Freedom.

To execute their responsibilities, the USAFE/LG staff was organized into cells to monitor and control CS activities. These control cells, operating in the absence of an established communication network, were in constant pursuit of timely and accurate data to support their C2 decisions. Each of the cells used some degree of manual data gathering and analysis. Additionally, they resorted to ad hoc reporting from the units. Their data gathering and analysis resulted in varying degrees of success with respect to data accuracy and timeliness. Similar cells for Civil Engineering and Services operated with the same issues.

During Operation Enduring Freedom, ACC made arrangements with CENTAF A-4 at the beginning of the engagement to act as the CENTAF A-4 Rear. Current doctrine calls for the NAF to be augmented and have a Rear function at the NAF site. In this case, ACC was in a better position and had greater resources to devote to the CENTAF A-4 Rear function. ACC co-located civil engineering (CE), logistics, and services to become the single focal point for CS actions. The A-4/7 Rear group was established as an around-the-clock operation with a colonel A-4/7 Combat Support director always on duty. The A-4/7 Rear reached back to ACC/LG Logistics Readiness Center (LRC) and ACC/CE Contingency Response Cell (CRC) for needed staffing. Again, this is not codified but could be a model for future operations. The MAJCOM could serve as the AFFOR Rear forces, consistent with initiatives at the MAJCOMs. PACAF has established a staffing package to support an Air Logistics Operations Center (ALOC) to provide staffing to support the Pacific Operations Support Center (POSC). The POSC is an AFFOR Rear-like organization, with co-located and grouped CS organizational functions. Maintenance, supply, fuels, and transportation are grouped to focus logistics support and reach back to the PACAF/LG LRC for depth. Similarly, CE, services, security, and base communications are grouped to provide installation support, with depth from functional staff cells. The ALOC would provide permanent staffing for an LRC-like organization.[12] Similar initiatives are taking place in USAFE, which has a theater aerospace support center that is organized in an AFFOR structure and has been charged to handle AFFOR functions.[13]

The permanent establishment of a cadre of support personnel for continuous contingency planning and execution could ease some of the augmentation issues. There would be fewer sites to augment and some of the NAF staffing could potentially be used to provide sources for the new organizations. Alternatively, the role of contingency planning could be a primary NAF task, with the NAF planners relocating to ACC, USAFE, and PACAF in contingencies. Further study and possibly additional manpower are needed.

The fact that different commodities fall under the responsibility of different organizations complicates CS resource assessment. Although commodities have different characteristics that may dictate that they be handled and managed in distinct ways, they need to be viewed from the perspective of how, in concert, they affect weapon system combat capability. Data are recorded in separate information systems, poli-

---

[12]BGen Pat Burns, February 26, 2002.

[13]USAFE Theater Air Support Center (UTASC) mission update brief, Col Forsythe, February 20, 2002.

cies and procedures vary for each of these organizations, and decisions are made on an individual commodity basis rather than from a comprehensive support perspective. This "stove-piping" of decisions affecting resource prioritization can lead to an imbalance between desired and actual capability and may not accurately reflect sortie production capability.

The operators, logisticians, and installations support personnel that we interviewed stated that the means by which CS information is made available to planners and operators is inadequately defined and inefficient.[14] Logistics and installation support personnel repeatedly expressed frustration over ever-increasing informal requests for information and tasks, attempts to bypass recognized channels, and uncertainty about whether the information they had was consistent with that in other CS-related offices. In addition, they pointed out that it is often unclear where and when logistics and installation support information, such as situation reports, should be transmitted. For example, during OEF, it was not uncommon in critical early weeks for deployed units to initiate reporting with their parent MAJCOM, leaving CENTAF Forward and/or ACC, as CENTAF Rear, out of the loop. As discussed earlier, these problems have a deleterious effect on overall system and operational efficiency.

An important step toward resolving these problems is to establish a C2 node template for CS, such as that shown in Table 5.3, with clearly defined responsibilities for each node. Appendix C details the information flows into these nodes, processes that take place within the nodes, and products that leave each node for other nodes.

The node template is a key element of the *TO-BE* operational architecture. The template will promote clearly delineated roles and responsibilities, process activities, and information flows assigned to each node. As contingencies evolve or dissolve within geographic theaters of operation, specific organizations will be designated to fulfill the responsibilities of one of the nodes. The template allows for variations in organizational assignments by theater, and may serve as a guide to configuring the C2 infrastructure, while retaining standardization of responsibilities. Along with the template, having standing C2 nodes could enable continuous CS execution planning and control for ongoing contingency and peacetime operations worldwide.

The need for standing CS organizations for C2 derives from the AEF environment. To respond to threats, AEF CS resources may need to be reallocated from one theater to another. Currently, some resources [such as theater-based munitions and war reserve materiel (WRM), intratheater distribution assets, and physical installations and operational infrastructure] are confined to individual theaters and are managed by theater-based organizations. This arrangement may remain effective, but the ability to relocate and reallocate to other AORs needs to be streamlined.[15] Other CS commodities are currently managed by units, but with the advent of CIRFs, some [e.g., LANTIRN pods, electronic warfare (EW) pods, engines] may need to be managed from a global perspective—moving limited assets quickly from one theater CIRF to

[14]Col Ed Groeninger, PACAF 502/CC, March 8, 2001; Lt Col Stephen Luxion, Hq CENTAF A-3/A-5, February 8, 2001; BGen Pat Burns, PACAF/CE and ACC/CE, February 26, 2002.

[15]See Feinberg et al. (2002).

another. Other examples of scarce resources that may need to be managed centrally include specialized equipment, spare parts, fuel, CONUS-based munitions, aerospace ground equipment (AGE), fuel mobility support equipment (FMSE), and consumables, as well as maintenance and intertheater distribution assets.

CS resource assessment and allocation management could be assigned to permanent organizational nodes dedicated to consumable and reparable resource moni-

Table 5.3

Air Force C2 Node Template for Combat Support

Hq Air Force
Global Integration Center (GIC)[a]
AFFOR
Air Operations Center CS element
AFFOR A-4 staff (and AFFOR A-7 staff if used)
Operations Support Center (OSC)[b] (theater/region)
Deployed Units
Wing Operations Center (WOC)
Combat Support Center (CSC)
Supporting Commands (force providers)
LRC or CSC
CRC
Deploying Units
WOC
Deployment Control Center (DCC)
Commodity Inventory Control Points (ICPs)
Munitions ICP
Construction materiel ICP
Specialized/heavy equipment ICP
Spares ICP
POL ICP
Bare base equipment ICP
Class II ICP (clothing, chemical gear, etc.)
Rations ICP
Medical materiel ICP
Personnel ICP
Vehicle ICP
Sources of Supply (depots, commercial suppliers, etc.)
Command Centers

[a]The GIC could be a virtual organization rather than a single physical organization. For example, a branch of the virtual GIC could be an analysis cell for combat fighter aircraft that is co-located with the ACC RSS for data access convenience. Similarly, another analysis cell for airlift and refueling aircraft could be co-located with the AMC RSS. Space and adequate computer links would be necessary for data access. The CSC, which oversees integration, could be another element.

[b]The theater/regional OSC would have AFFOR Rear responsibilities in support of multiple COMAFFORs within a single combatant command theater or across multiple theaters.

toring, prioritization, and reconfiguration. Additionally, having a standing integration function for CS resource management will facilitate the incorporation of relevant data into capability assessments and raise the visibility of these assessments in the eyes of the operational community. Table 5.4 reflects the roles and responsibilities of the organizational nodes in the template and addresses the roles of parallel nodes in Joint organizations. Appendix C contains more detail on the nodes and their responsibilities as well as more detail on the information needed, processing, and information produced and sent to other nodes.

To coordinate resource-level management, the Air Force could put a CS theater presence in the Operations Support Center. The OSC will act as regional hub for monitoring, prioritizing, and allocating theater-level CS resources, and be responsible for mission support,[16] base infrastructure support, and establishing movement requirements within the theater. The OSC will be the theater integrator for commodities managed by Inventory Control Points (discussed below). To be effective, the OSC must have complete visibility of theater resources and the authority to reconfigure them. It will receive commodity-specific information from ICPs and make integrated capability assessments (both sortie production and base) and report those assessments to the CS personnel supporting JAOP/MAAP/ATO production in the AOC. In this role, the OSC will make allocation decisions in the face of competing demands for resources. Finally, it must work closely with the joint-service forces community to ensure that resources are allocated in accordance with global priorities. Impact analyses will justify demands for critical resources in competition with other theaters. The OSC could incorporate mission, base infrastructure, and movement capability assessments into operational plans. It would support the deployed AFFOR A-4 staff during a contingency, minimizing the number of personnel required to deploy forward. It would also alleviate problems associated with an undermanned NAF staff currently trying to perform the functions listed above as well as their roles under the unified command structure.

Inventory Control Points could manage supply of resources to the MAJCOMs—essential for the distribution of critical resources such as munitions and spares. For example, spares should be managed along weapon system lines by an ICP run by Air Force Materiel Command (AFMC). The standing C2 node at AFMC would manage spares along the continuum of operations, with immediate access to both the data and analytical tools needed to assess capability and manage distribution of resources to MAJCOMs and theaters under direction from an integrating function. The spares ICP would monitor resource inventory levels, locations, and movement information,

---

[16]With today's communications and computer technology, it can be argued that analysis cells of the virtual GIC at ACC, AMC, and Space Command (SPACECOM) could assess and provide worldwide support for combat, strategic lift and tanker, and space-related weapon systems with Air Force–wide integration provided by the CSC. This would reduce OSC responsibilities to providing beddown support and transportation priorities among sites within the theater. Doctrine currently assigns the combatant commander support responsibility for forces over which he has operational control (OPCON). This was not followed during Operation Enduring Freedom when AMC retained support responsibilities for KC-10s and KC-135s, except those assigned to Thumrait and Al Dhafra air bases. Mission capable rates for units that were engaged in the theater but remained under AMC support control were higher than those that were supported by the combatant commander. This doctrine needs to be revisited. We have, however, assigned the assessment and control function to the theater OSCs in this report.

Table 5.4

Nodes and Responsibilities

C2 Node	Role/Responsibility
Joint Staff	
Logistics Readiness Center	Supply/demand arbitration across Combatant Commands
Combatant Command	
Combatant Command Logistics Readiness Center	Combatant Command logistics guidance and COA analysis
Joint Movement Center (JMC)	Combatant Command transportation supply/demand arbitration
Joint Petroleum Office (JPO)	Combatant Command POL supply/demand arbitration
Joint Facilities Utilization Board	Combatant Command facilities/real estate supply/demand arbitration
Joint Materiel Priorities & Allocation Board	Combatant Command materiel supply/demand arbitration
Joint Mortuary Affairs Board	Combatant Command mortuary affairs management
Theater Patient Movement Requirements Center	Combatant Command medical patient movement prioritization
Joint Civil/Military Engineering Board	Theater engineering supply/demand arbitration
JTF	
JTF J-4 & Logistics Readiness Center	JTF logistics guidance
JTF J-7 & Contingency Response Cell (if used)	Supply/demand arbitration within JTF among service components
	JTF installations support guidance
JFACC	
Joint Air Operations Center CS Representatives	JAOP/MAAP/ATO production support
JFACC staff logisticians	JFACC logistics guidance
JFACC installations support	JFACC installations support guidance
Air Force	
Combat Support Center	Monitor operations
	Represent Air Force CS interest to Joint Staff
Global Integration Center	Integrated weapon system assessments
	Critical resource supply/demand arbitration across AFFORs

Table 5.4—continued

C2 Node	Role/Responsibility
AFFOR	
Air Operations Center CS element	Air Campaign Plan/MAAP/ATO production support
AFFOR A-4 Staff and AFFOR A-7 Staff (if used)	Site surveys/beddown planning
OSC (theater/region) (can support multiple AFFORs)	Liaison with AOC CS element
	Mission/sortie capability assessments
	Beddown/infrastructure assessment
	Aerospace Expeditionary Task Force (ASETF) force structure support requirements
	Supply/demand arbitration within ASETF among AEFs/bases
	Theater distribution requirements planning
	Force closure analysis
	Liaison with Air Mobility Division in AOC
	Liaison with theater TRANSCOM node
Deployed Units	
Wing Operations Center	Disseminate unit tasking
	Report unit status
Combat Support Center	Monitor and report performance and inventory status
Supporting Commands (Force and Sustainment Providers)	
Logistics Readiness Center/CSC	Monitor unit deployments
Contingency Response Cell	Allocate resources to resolve deploying unit shortfalls
Deploying Units	
Wing Operations Center	Report unit status
	Disseminate unit tasking
Deployment Control Center	Plan and execute wing deployment
	Report status of deployment
Commodity Inventory Control Points (ICPs)	
Munitions ICP	Monitor resource levels
	Assess capability
	Allocate resources in accordance with theater and global priorities

Table 5.4—continued

C2 Node	Role/Responsibility
Construction materiel ICP	
Specialized/heavy equipment ICP	
Spare ICP[a]	
POL ICP	
Bare base equipment ICP	
Class III CP (clothing, chemical gear, etc.)	
Rations ICP	
Medical materiel ICP	
Personnel ICP	
Vehicle ICP	
Sources of Supply (Depots, Commercial Suppliers, etc.)	
Command Centers	Monitor production performance and report capacity

[a]Global resource allocation decisions for spare parts are coordinated with the Global Integration Center.

and use these data to assess contractor and depot capabilities to meet throughput requirements.

In addition to the commodity ICPs and the theater OSC, a Global Integration Center is needed, as briefly discussed above. It could be a virtual organization with analysis cells co-located with the ACC, AMC, and SPACECOM RSSs to assess weapon system capabilities, and should have responsibility for providing integrated weapon system assessments across commodities both in peacetime and wartime. An Air Force–level cell (possibly the Air Force CSC) could integrate assessments that support allocation decisions when multiple theaters are competing for the same resources and could serve as the Air Force voice to the Joint Staff in any arbitration across services. With the global nature of AEFs and worldwide commitments, other commodities should be considered for management in the same manner.

At both the OSCs and the GIC, individual resource prioritization will be guided by a common set of rules: given a required operational capability, the OSC will calculate the CS resources needed to meet it. Multiple ways to achieve the same goals will be considered in resource prioritization. Resources will then be assessed and allocated to meet the operational capability requirements set at higher levels [e.g., the National Defense Strategy and Joint Chiefs of Staff (JCS)]. These resources will thus be allocated according to the need for an overall level of operational capability rather than on an individual commodity basis.

Based on these assessments and allocations, the ICPs will direct purchases, repair operations, and distribution of components and spares, coordinate with combatant commanders and the joint-services community for intertheater airlift and direct the distribution of resources among theaters, and provide commodity support capabilities analysis and assessment to GICs. Theater OSCs will advise of infrastructure capabilities, needed resources to implement plans, and the consequences of not improving capabilities. The theater joint-services command can then prioritize needs and advise the Joint Staff and others of theater capabilities and issues. Ongoing capability assessments generated by the GIC and OSCs will be incorporated into a theater's operational planning processes executed by CS liaisons in the AOC.

In this construct, distribution decisions consistent with operational priorities will be made at the lowest possible level. Elevation of the decision to a higher authority can be triggered by a number of factors, including constraints beyond the capacity of a lower echelon to solve and competition for resources extending beyond the decision authority of the lower level. Table 5.5 is an example of how these triggers might be established.

This organizational structure offers three improvements over the existing one. First, it enables prioritization and allocation based on operational capability assessments. Rather than simply giving the highest priority to all requests from a particular

Table 5.5

Resource Distribution Decision Triggers

Decision Level	Decision Authority	Decision Elevation Trigger	Elevation Level
		Ammunition Example	
Ammo ICP	Allocate munitions to an AFFOR in accordance with established priorities to meet planned requirements	Threshold breech driven by demand from multiple AFFORs *within single theater*	OSC
OSC	Allocate munitions within a single theater	Threshold breech driven by demand from multiple AFFORs *from different theaters*	GIC
GIC	Allocate munitions to AFFORs in different theaters	Resource competition resulting in capability degradation of one theater vs. another theater	SECDEF
		Spare Parts Example	
Shop	Allocate repair resources to meet planned requirements	Threshold breech crosses shop-allocated resources	Air Logistics Center (ALC) or Purchasing and Supply Chain Manager (PSCM)
PSCM	Buy/distribute/repair	Threshold breech by weapon system or commodity	Virtual Inventory Control Point or GIC
OSC (regional)	Allocate among competing requirements within a theater	Threshold breech driven by demand from multiple AFFORs *from different theaters*	GIC
GIC	Buy/distribute/repair priorities across weapon systems and commodities Resource competition among AFFORs across theaters	Resource competition resulting in capability degradation of one theater vs. another theater	SECDEF

location, the status of each location is synthesized from several global information control points. Capabilities are estimated in the context of theater and global priorities, and resources are allocated accordingly. Distribution of CS capabilities is more informed, resources can be moved before requests are made, and filling emergency requests is easier. The second distinction is that this structure considers the complete spectrum of CS resources. Each resource influences operational capability in some way, and hence must be prioritized and allocated in conjunction with the others. By centralizing CS capability assessments, capability becomes a commodity that can be managed like any other, with a single set of decisionmakers. Although this management is ultimately broken down into the movement of individual resources, the resources are not managed individually but rather in an integrated manner. The third strength is that by establishing nodes for designated tasks, the structure is a consistent framework for decisionmaking throughout all phases of operations. Because the standing nodes are devoted to the monitoring, prioritization, and reconfiguration of all CS resources, they are equally capable of addressing long-term weapon system development considerations, training, or crisis action planning and execution.

These responsibilities can be performed by organizations in different theaters, but the grouping of the tasks, the information required to complete them, and the products resulting from each task should not change from one theater to the next. Table 5.6 reflects a notional assignment of organizations to nodes in the template, creating a theater-specific view of the C2 architecture for CS functions. Assigning organizations to perform each task will better define the communication network and roles that each augmenter needs to train for. This will result in improved training programs and better-trained personnel in wartime positions. With better-trained staff, planning organizations will be better able to assess operational capability, monitor deployments, and respond to changes in an ongoing campaign.

TRAINING AND EDUCATION

To move toward the *TO-BE* decisionmaking process, the Air Force must make several changes in its training and education programs. Table 5.7 lists shortfalls in current training and changes required to enable the *TO-BE* concept.

As indicated earlier, the absence of a core process and well-documented C2 operational architecture baseline for CS contributes to the shortfall in training and education. For example, ineffective communication between operation and support planners can be attributed to the fact that CS personnel typically do not have experience in both logistics and installation support functions and are not effectively taught their role in the context of operational planning. As a result, they do not develop metrics appropriate for communicating with operators, the Joint community, or other members of the diverse support chain.

Similarly, because operators lack an understanding of how CS contributes to and enables operational capabilities, they often set strategies without sufficient CS input, which leads to the difficulties discussed earlier. Exercises often lack CS realism,

Table 5.6

C2 Nodes and Theater Organization Notional Alignments (CS Elements Only)

Combat Support C2 Node	Notional Alignment of Organizations by Theater		
	USAFE	CENTAF	PACAF
Joint Staff			
Logistics Readiness Center	Joint Staff LRC	Joint Staff LRC	Joint Staff LRC
JTF			
JTF J-4 and Logistics Readiness Center[a]	Appointed JTF J-4/LRC	Appointed JTF J-4/LRC	Appointed JTF J-4/LRC
JFACC			
Joint AOC CS representatives	Appointed JFACC	Appointed JFACC	Appointed JFACC
JFACC staff logisticians	32 Air Operations Group (AOG) logisticians	9AF logisticians	11AF/13AF logisticians 607 ASUS
Air Force			
Global Integration Center[b]	AF virtual GIC	AF virtual GIC	AF virtual GIC
AFFOR			
Air Operations Center (AOC) CS element	32 AOG logisticians	9AF logisticians	11AF/13AF logisticians 607 ASUS
AFFOR A-4 staff	16AF/3AF logisticians	9AF logisticians	11AF/13AF logisticians 607 ASUS
Operations Support Center (OSC) (theater/region) (can support multiple AFFORs)	USAFE Theater Aerospace Support Center (USAFE TASC)	CONUS OSC (ACC RSS expanded)	PACAF OSC (POSC)
Deployed units			
Wing Operations Center (WOC)	Deployed units' WOC	Deployed units' WOC	Deployed units' WOC
Combat Support Center	Deployed units' LG staff	Deployed units' LG staff	Deployed units' LG staff
Supporting Commands (Force Providers)			
Logistics Readiness Center/CSC	Force providing MAJCOM LG staff	Force providing MAJCOM LG staff	Force providing MAJCOM LG staff
Deploying Units			
Wing Operations Center (WOC)	Deploying units' WOC	Deploying units' WOC	Deploying units' WOC
Deployment Control Center (DCC)	Deploying units' LG staff	Deploying units' LG staff	Deploying units' LG staff
Commodity Inventory Control Points (ICPs)			
Munitions ICP	OO-ALC/WM USAFE TACP	OO-ALC/WM	OO-ALC/WM PACAF TACP
Spares ICP	AFMC/DLA	AFMC/DLA	AFMC/DLA

Table 5.6—continued

Combat Support C2 Node	Notional Alignment of Organizations by Theater		
	USAFE	CENTAF	PACAF
POL ICP	DESC	DESC	DESC
	DESC—Europe	DESC—Middle East	DESC—Pacific
Rations ICP	DLA	DLA	DLA

NOTE: TACP = Theater Ammunition Control Point; DLA = Defense Logistics Agency; OO-ALC/WM = Ogden Air Logistics Center/Air-to-Surface Munitions Directorate; DESC = Defense Energy Support Center.

[a]The J-4, A-4, and LRC installation responsibilities can be split off into a J-7, A-7, and Contingency Response Cell. In this table, the J-4, A-4, and LRC has responsibility for both logistics and installations.

[b]GIC could have cells at AMC, ACC, AFMC, and Air Staff or some combination of them.

Table 5.7

Training Shortfalls and Solutions

Training Shortfall	Proposed Solution
Most operations and CS training focused on wing-level skills, not operational-level skills	Develop CS course curriculum for C2
Little training for CS personnel on operations C2, and for operations personnel on CSC2	Expand the role of CS in wargames/exercises
Little training on communicating and operating with the joint-services community	Take advantage of joint-services logistics wargames (e.g., FLOW) to evaluate new concepts
CS participation in exercises and wargames not accurately addressing the execution planning process	Incorporate C2 gates in CS officer and enlisted career development
Few training opportunities	Develop C2 job performance aids for CS

which carries over to real-world contingencies in which operational planners generally do not consider CS issues until well into the planning process. OEF experience bears this out—operational forces arriving well in advance of their combat support were mission-hampered and under severe beddown living conditions.

This lack of awareness of each other's roles and processes, and inability to communicate between operations, logistics, and installations support becomes particularly evident in COA development during crisis action planning. Combat support personnel describe their capabilities in terms of the amount of fuel, munitions, and spare parts available. Operations planners are more interested in assessments of logistics and installations support infrastructure that relate CS resources to aircraft basing and sortie production. With proper training and education, this information could be incorporated into strategy at a much earlier point, but CS planners do not know how nor do they have the tools to provide it.[17]

[17]Lt Col Stephen Luxion, Hq CENTAF A-3/A-5, February 8, 2001. Col Duane Jones, CENTAF Forward A-4/A-7, February 14, 2002.

During our site visits and interviews, we learned that many warfighting staff members are not adequately trained in their management roles. Most staff are assigned to regional CS roles from the wing-level and have little or no experience in the diversity of CS resource management at a regional level. Little formal training is available to develop such skills. In fact, few opportunities for C2 training exist, leaving both operations and CS personnel to learn their responsibilities through on-the-job training (OJT). OJT is problematic because manning for many command-level support functions at the NAFs is limited. Some OJT is necessary, but without supplemental information it can reinforce bad practices and bypass issues that are not raised on a day-to-day basis. Examples of skills that are not formally trained include managing the regional supply chain, non-unit sustainment/resupply resources, and theater-owned resources, as well as administering interactions between bases, MAJCOMs, headquarters, joint-services forces, and the operations community. That, coupled with the absence of detailed policy, leaves many warfighting staff members and augmenters with little help in understanding how to execute their responsibilities.

Training and education shortfalls can be remedied. Training can be improved through the development of a CS curriculum, which can be incorporated into existing and upcoming training courses. The JAC2C (Joint Aerospace C2 Course) at Hurlburt Air Force Base is the joint-services introductory course for basic AOC processes. This course can be expanded to include elements of operational-level CS planning and execution. For operations personnel, JAC2C is mandatory training for designated assignments. Combat support personnel should be encouraged and funded to attend these courses with the same priority. Additionally, the Chief of Staff Logistics Review (CLR) initiative to implement a logistics officer weapons school should include a link to the CS execution planning and control curriculum.

As a longer-term goal, the CS curriculum should be incorporated into new courses. In addition to JAC2C, courses should train on:

- CS doctrine, policy, and guidance
- AFFOR and AOC CS processes
- Weapon system and infrastructure capability assessments, to incorporate CS resource levels and other metrics into both theater and global capability measures
- New decision support tools, as they are developed and implemented.

Exercises and wargames should include more CS issues and be funded to educate both operators and CS planners on their respective roles and the role of CS resources in campaign planning.

Career-path planning for CS personnel might include assignment to warfighting command-level positions in supply, transportation, logistics plans, CE, services, etc., with the intent of creating senior CS personnel with the skills needed to fill AFFOR A-4 (and A-7, if used) and combatant command staff CS positions. The Developing Aerospace Leaders (DAL) initiative offers a good opportunity to establish the breadth and depth of experience future CS leaders need.[18] Developing this path requires more closely identifying and monitoring the track for CS positions. Air Force leadership can identify which existing (and forthcoming) courses better prepare their CS students for C2 roles. Those CS officers with a strong C2 background can be groomed for leadership positions, and those with weaker backgrounds can be given supplemental training.

A basic element of the *TO-BE* concept is the feedback loop, which enables CS input to affect operations planning. For the feedback loop to be most effective, CS personnel must understand air campaign planning and aerospace force capabilities. For example, what issues factor into planning different phases of the air campaign? What factors drive weapon systems and preferred munitions selection? What other weapons can provide similar effects? The CS planner of tomorrow, working side-by-side with operations planners in an integrated planning process, must be able to answer these questions. Changes to training and education should equip CS personnel to translate logistics and installation support resources to operational capabilities.

Finally, decision support tools and job performance aids should complement formal courses and exercises. Air Force Tactics, Techniques, and Procedure (AFTTP) publications should provide desktop guidance on the use of tools, delineate the roles and responsibilities of each organization, and provide insight into the decisionmaking process. Similarly, web-based process guides can assist wartime planners on their C2 responsibilities and their roles in execution planning.

Training is a critical aspect of the link between CS and operational planning. Educating both CS and operations personnel about their roles in the context of campaign planning will enable more effective communication and facilitate the integrated decisionmaking process in the *TO-BE* architecture.

INFORMATION SYSTEMS AND DECISION SUPPORT

Several aspects of both the *AS-IS* and the *TO-BE* CS planning processes would benefit from enhancement or development of information systems and decision support tools. Table 5.8 identifies several shortfalls in the current process that could be remedied with decision support tools and an improved infrastructure.

CS resource assessment, prioritization, and reconfiguration has been emphasized throughout our description of the *TO-BE* architecture and is illustrated in detail in Appendix C. Existing systems are unable, for several reasons, to support these capabilities,[19] primarily because of the lack of uniformity among systems. Because CS resources have been managed by "stove-pipes" and funded by commodity, with different organizations having commodity management responsibility, corresponding

---

[18]MGen Chuck Link (USAF, ret.), DAL Video, January 2002.

[19]Lt Col Stephen Luxion, Hq CENTAF A-3/A-5, February 8, 2001; Hq AFMC LGXX, February 21, 2001; Mr. Van Hazel, 7th Air Force operations research analyst, December 10, 2001; Major Parker Northrup, 7th Air Force Air Operations Group, December 10, 2001; Major Steen, PACAF/XPXX, December 17, 2001; Lt Col Levault, 13th AF/A3/5, December 13, 2001.

Table 5.8

Decision Support Shortfalls and Solutions

Tools/Systems Shortfall	Proposed Solution
Tools needed to:	Develop tools to provide required capabilities
Relate operational plans to CS requirements	Focus integration efforts on global implementation of a few selected tools
Convert CS resource levels to operational capabilities	Standardize tools and systems for consistent integration
Aggregate capability assessments to a theater or global scale	
Conduct capability assessment and aggregate them on a theater or global scale	
Conduct tradeoff analyses of operational, support, and strategy options	
Inability to access data on a timely basis	
Proliferation of tools and systems has resulted in marginal success in fielding capabilities	

information systems have been developed and implemented independently among the organizations. The result is a myriad of independent systems with little ability to share data or interface with other systems.[20] Thus, although these systems allow individual commodity data to be recorded and monitored, they do not facilitate the integration of the data for comprehensive CS resource monitoring and capability assessments. Furthermore, with such a proliferation of systems, data in each are updated only sporadically,[21] and update status and data reliability are often unknown to users.

Existing information systems also lack robustness. Reliable recording of time-sensitive and often classified data within a globally distributed mobile organization like the Air Force is inherently challenging.[22] For example, logistics planning factors, which govern the translation of operational plans to CS resource requirements, are updated only every few years.[23] Similarly, base/host nation infrastructure capacity is only updated on an as-needed or contingency-driven basis.[24] These factors result in CS plans that are not reliable. In addition, CS planners may not be aware of tools available to estimate CS requirements.[25]

---

[20]Hq CENTAF A-4 Supply, February 7, 2001.

[21]Interviews, Joint Staff J-4, Pentagon, February 23, 2001.

[22]Ed Kowzlowski et al., Hq AFMC, May 2, 2001.

[23]Walter Busby et al., Hq CENTAF A-4, February 8, 2001.

[24]Col Tom Ryburn et al., Hq CENTAF A-7, February 15, 2002.

[25]In the AWOS, USAFE CS planners were not aware of LOGSAFE, a tool to estimate resupply transportation requirements. See Feinberg et al. (2002). During the early months of OEF, AMC CS planners were unaware of ACC and PACAF GeoReach remote base imaging and mapping capability. Ryburn et al., February 15, 2002.

Systems are needed to constantly monitor CS capacity, resource inventory, and process performance levels, and tools are needed to convert operational plans and status to CS resource requirements and resource levels and then into operational capabilities. Tools are also needed to inform maintenance workload decisions by expressing infrastructure status in terms of operational capabilities and estimating resupply, beddown, and associated sustainment requirements. These tools will enable the Air Force to more accurately express their resupply and sustainment needs. Finally, tools are needed to aid beddown decisions. Some of these requirements can be supported by integrating and modifying existing systems, whereas others will require new system development.

A thorough evaluation should consider all decision support tools for a particular function, with implementation focused on a smaller set of tools worldwide. This will reduce the number of systems and training programs required for each planning function and permit an efficient transfer of information.

New tools should be built on a systems infrastructure that can rapidly transfer information to maintenance facilities, inventory control points, OSCs, and other key CS nodes, as well as the AOC and all relevant operational nodes. This infrastructure will maximize the productivity of new tools and allow them to interface with joint-service systems. Air Force actions can then be framed in the context of a joint-service campaign with information disseminated on a timely basis.

The effects of improved information systems and decision support tools will be felt throughout the *TO-BE* process. Properly integrating information from these tools will greatly reduce the chances of needing to revise a plan in midstream, allow a faster transition to war and better-informed decisions, and facilitate change when necessary.

Enhancement of information systems and decision support tools is a challenging and difficult task in any organization, but it is particularly challenging in the Air Force CS area because of the new C2 functions that need to be supported. The value of each additional capability will need to be considered as well as its cost. The Air Force may consider seeking external advice on how to best address this issue. The architecture presented here provides a view of the processes and functions that must be developed to better develop CS planning and execution responsibilities across the spectrum of operations. If adopted, this architecture can keep information system and decision support system development on target.

Chapter Six

SUMMARY AND CONCLUSIONS

Closing the gap between the *AS-IS* and *TO-BE* CS execution planning and control systems requires significant changes to current doctrine and policy. Both doctrine and policy should emphasize the CS execution planning and control roles; describe the relevant objectives, functions, and activities of such a system; and define organizations to carry out the functions and activities.

Once doctrine and policy are in place, current processes can be revised to integrate combat support and operations planning as well as combatant command and joint-services planning, allocate resources according to required capabilities, and create a closed loop between planning and execution functions, thus enabling better informed plans as a campaign continues.

Standing C2 organizations for CS, with clear chains of communication and well-defined responsibilities, could facilitate CS planning and execution. Combat support could better become part of operations planning with the creation of a support organization to collaborate with the AOC in developing a single, integrated plan. Distribution and resource management could be improved by a network of commodity-specific resource managers and an integration function to make allocation decisions when faced with competition for resources.

Changes to the *AS-IS* system should be reinforced with training and exercises. Developing a C2 course curriculum and incorporating C2 milestones into CS career development programs will better train both officers and enlisted personnel for their CS roles. Additionally, expanding the role of CS in wargames and exercises should demonstrate its importance in contingency operations and help ensure that operational and strategic personnel consider CS issues in plan development.

The changes also require different information flows and development of decision support tools, implemented on a robust information systems infrastructure. These tools should focus on execution planning and tradeoff analyses, and include operational and support metrics, global and theater capability assessments, and the efficiency of CS processes.

Appendix A
INTERVIEW LIST

Command	Rank	Name	Group/Organization
HQ PACAF	Col	Ed Groeninger	502 AOG/CC
	Col	Dave Smith	LGX
		Elaine Ayers	LGX
		Randy Stewart	LGX
	Capt	Fraser	SC Group
	BGen	A. B. Morrill	PACAF/LG
	Maj	Steen	XPXX
	MSgt	Ouzto	PACAF/LGWX
		Gregory Osbun	
	CMSgt	Ivy	607th ASUS
	CMSgt	Pellegrino	607th ASUS
	Capt	Katowich	607th ASUS
	Lt Col	Perry	607 DDK
		Harold Van Hazel	607th ASUS
	Lt Col	Foote	607th ASUS
	Capt	Froemke	607th ASUS
	Capt	Shaser	607th ASUS
	Maj	Sabo	607th ASUS
7th AF	Col	Russ Grunch	A-4
	Maj	Parker Northrup	A-5
HQ ACC	MGen	Don Wetekan	LG & A-4
	BGen	Pat Burns	CE & A-7
	BGen	Mike Collings	ACC/LG
	Col	Huck Robinson	LGX
	Capt	Jennifer Murphy	XP
AC2ISR	Col	Bill McGill	ACS Division Director
	Col	Peaches Cavanaugh	ACS Dep Director
	CMSgt	Pete Conrad	RSS
	SMSgt	Moore	RSS
ACC	Capt	Dave Barna	RSS
	Maj	John Beecy	RSS
		Les Parnacott	RSS
		Tony Mattox	RSS

Command	Rank	Name	Group/Organization
	CMSgt	Pete Conrad	RSS
	SMSgt	Moore	RSS
Joint Staff	Col	Ed Hatch	J-4
	Col	Jack Welsh	J-4
	Lt Col	Giroux	J-4
	Lt Col	Salesses	J-4
	Lt Col	A. Ray Myers	J-4
	Lt Col	Brent Baker	J-4
	Lt Col	Carl Puntureri	J-4
	SMSgt	Joseph Hudgins	Fuels
USAFE	Col	Maury Forsyth	UTASC/CC
	Maj	Maria Garcia	LGXP
	Lt Col	Thomas Lisk	US ANG
	Lt Col	Lennie Edwards	32 AOS
	Lt	Nate Harris	32 AOS
	Maj	William Ward	32 AOS
	Lt Col	Edward Appler	86 CRG
	Lt Col	Michael Marra	86 CRG
	Capt	Strassberger	86 CRG
	SMSgt	Grady Huffman	86 CRG
	SMSgt	John Made	86 CRG
	MSgt	William Maus	86 CRG
	TSgt	Paulo DaSilva	86 CRG
	SMSgt	Strickland	RSS
	MSgt	Olney	RSS
	MSgt	Miller	RSS
	Capt	Cotto	RSS
	Lt Col	Sharon Holmes	AMOCC
	Maj	David Meyer	AMOCC
	Maj	Hiawatha Newton	AMOCC
	Maj	Todd Coats	AMOCC
	Lt Col	Larry Hudson	LG
	Capt	Darrel Cunningham	LG
	BGen	Art Rooney	LG
	Capt	Dory Traversa	LG
	Capt	Jeff Burrell	LG
	Capt	Patrick Walker	LG
	Maj	Glen Slotness	LG
	Maj	Tom Schneider	LG
		LG staff	LG
		Dave Parmley	LG
		James Kibbee	LG
		John Coon	LG
	Lt Col	Dale Coliaianni	LGTV
	Capt	Edward Peterson	LGTV
	Capt	Hearn	LGTV
		Klaus Waismantel	LGTV

Command	Rank	Name	Group/Organization
	Col	Mark Jones	LGX
	Lt Col	Brad Silver	RSS Commander
	Lt Col	Frederick	RSS Commander
HQ USAF	BGen	Darryl Scott	AQC
CENTAF	Col	Duane Jones	A-4
	Col	Tom Ryburn	A-7
	Lt Col	Forrester	A-4 Supply
	Lt Col	Papucci	A-4 Transportation
	Lt Col	Bill Doneth	A-3 Operational Plans
	Lt Col	Stephen Luxion	A-3/A-5 DOXP
	Lt Col	Darc Nelson	A-7 CE
	Maj	Nankervis	Log Plans
	Maj	Donald Thibeault	Log Plans
	Maj	Eason	Transportation
	CMSgt	Tate	Fuels
	SMSgt	Charles Swaggart	Ammo
	MSgt	Gulledge	Fuels
	MSgt	Alex Ritchey	LGXX
	MSgt	John Parrot	Maintenance
	TSgt	Carlos Cuevas	Weapons Mgr
		Walter Busby	LGXX
13th AF	Col	Carol King	A-4
	Lt Col	Levault	A-3/A-5
	Col	John S. Jaczinski III	Vice Commander 13th AF
PACOM	Gen	Geehan	PACOMJ4
	Col	Cooper	PACOM DJ4
AFMC		George Zeck	LGIP
		Larry Fortner	LGX
		Brad Baskin	LGXX
		Jim Weeks	LGXX
		John Frabotta	LGXX
		Tom Jenkins	LGXX
		Ed Kozlowski	XPAO
	Lt Col	Tom Fritz	XPAO
		Curt Neumann	XPS
		Rich Moore	XPS
AMC	BGen	Peter Hennessey	AMC/LG

Appendix B

AS-IS CSC2 DETAILED PROCESS FLOW MODEL

Figure B.1 is a detailed representation of the Air Force's current, or *AS-IS*, CSC2 operational architecture, which is discussed in Chapter Three of this report. The process flow model was constructed from interviews with subject matter experts (see Appendix A for a list of interview participants); reviews of Air Force and joint-services doctrine, manuals, instructions, and CONOPs; analyses of lessons learned from the Air War Over Serbia; and insights from previous studies such as the AFFOR baselining exercise conducted by the Aerospace Command and Control, Intelligence, Surveillance, and Reconnaissance Center (AC2ISRC).

The figure emphasizes processes and information flows. Each box represents a process, with the arrows connecting the boxes representing products and other types of information inputs and outputs to the processes. In general, the processes flow sequentially from left to right, from peacetime training to redeployment. The iterative nature of many processes is represented by reverse arrows, which show progressively refined information inputs and outputs. Each process box is labeled with a primary reference indicating its source.

The vertical axis is a generalized view of the organizational structure of the Air Force and joint-services community, so that the height of a process in the diagram reflects the approximate organizational level at which it occurs. Although some processes are nominally associated with specific organizations, no attempt has been made to make such assignments systematically.

This graphic readily illustrates the relationships between processes and records the information flows between them. It also helps identify critical points in the system, such as processes that receive insufficient inputs, processes that require numerous inputs, and products that are required for numerous processes. Such critical points were used to help identify shortfalls in the *AS-IS* system and guide the development of the *TO-BE* concept.

TO-BE CS EXECUTION PLANNING AND COMBAT DETAILED PROCESS FLOW MODEL

DIAGRAM AND DATABASE

Figure C.1 is a process diagram of the *TO-BE* CS execution planning and control operational architecture. Each primary activity is depicted in greater detail (tasks and information flows) in the supporting database. The activities are distinguished by operational phase and organizational node. The phases include readiness, crisis action planning, deployment, employment, and sustainment. The eight organizational nodes range from the President, Secretary of Defense (SECDEF), and associated high-level joint-service organizations to generic sources of supply (SOSs) for individual commodities. As described in Chapter Four and discussed further here, assigning tasks to nodes rather than to individual organizations allows for standardized roles and responsibilities across different theaters even if the organization occupying that node is theater dependent.

The diagram is drawn to emphasize processes and information flows. Each box represents an activity, with the arrows connecting the boxes representing products and other types of information inputs and outputs to that activity. Although the processes generally flow from left to right, the activities are not necessarily performed sequentially. For example, a node may show tasks in both the readiness and crisis action planning phases. The iterative nature of many processes is shown by feedback loops, representing progressively refined information inputs and outputs.

Note that in several places multiple activities are contained within a larger activity that may span more than one organizational node. This notation conveys the importance of cooperation between different nodes for certain activities. An example is the collaboration between the JFACC/AFFOR and OSC in the various planning stages. As discussed below, this is a major component of the *TO-BE* architecture.

An HTML version of the *TO-BE* process map can be found on the CD enclosed with this report. Individual activities and arrows on the diagram are linked to the database, allowing the user to explore the relationships among activities, tasks, phases, and nodes. Information inputs and outputs between activities can be viewed by dragging the mouse over the connecting arrows. Clicking on an activity on the diagram brings up a table of tasks associated with that activity, as well as a listing of the individual information flows into and out of it. Clicking on an organizational node heading brings up all of the activities, tasks, and information flows associated with that node throughout the different phases. Clicking on an operational phase heading will similarly bring up all of the activities, tasks, and information flows associated with that phase across the different nodes. Finally, the user may select to view only those data associated with a particular combination of phase and node.

The most important modifications to the CS execution planning and control architecture are in theater-level combat support (as conducted by the AFFOR and the OSC) and inventory management (the focus of the Inventory Control Points and Global Integration Center). Consequently, these organizational nodes are portrayed here in somewhat greater detail relative to other nodes in which few changes were made. For example, no modifications were proposed for the SECDEF/JCS/combatant command/JTF level, so this node is not included in the database. The SOS node is similarly excluded; any important decisionmaking regarding supply is made at the ICP and GIC nodes.

NEW ARCHITECTURAL COMPONENTS TO IMPLEMENT *TO-BE* CONCEPT

Chapter Three outlines several shortcomings in the *AS-IS* system identified by comparing analysis and documentation of the *AS-IS* system with the *TO-BE* concept derived from AEF goals. *AS-IS* system aspects that hinder effective implementation of the *TO-BE* concept were defined as shortcomings. Several of the changes proposed to help transform the CS system to conform with the *TO-BE* concept are reflected in Figure C.1 and the accompanying database. We next discuss these changes and how they address the shortcomings. Chapter Four describes the individual activities and organizations shown in Figure C.1.

Increased Integration of CS Input Into Operational Planning

Beginning in the readiness phase (which includes deliberate planning) and continuing through the crisis action planning phase, the theater or regional operational planning activities receive explicit input from the CS community (see activities A31, B31, and B32 in Figure C.1). In each case, the JFACC/AFFOR-level planning process is guided by both strategic input from SECDEF and the joint-services staff as well as resource and capability input from the CS community (i.e., the OSC).

This differs substantially from the *AS-IS* system, where operational plans are developed largely independently of CS input. Plans are subsequently sent to the CS community where support plans must be developed or the plan is rejected as infeasible from a CS perspective. As discussed in the main text, this serial approach can result in prolonged development of unsupportable plans that may require major restructuring when CS is factored in. An integrated planning process would contribute substantially to COA assessment, thereby focusing efforts on feasible COAs early in the planning process.

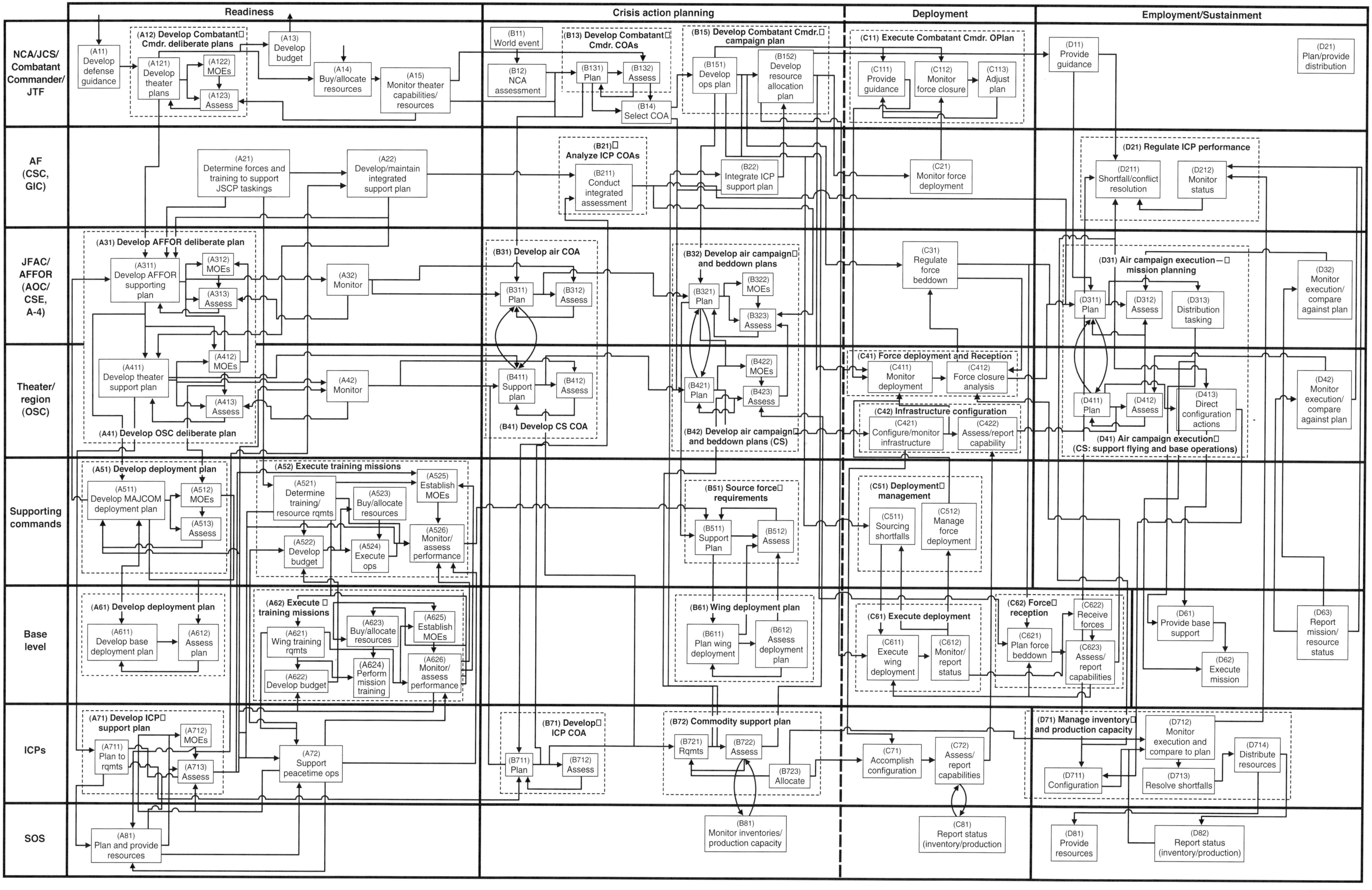

Figure C.1—TO-BE CSC2 Process Map

Feedback Loops to Reconfigure the CS Infrastructure in Response to Changing Demands or Capabilities

Feedback loops influencing CS activities occur at several points in Figure C.1. The most important example is in the employment/sustainment phase, in the set of activities leading into and out of activity D21. This "loop" conveys regulation of CS activities relative to performance criteria and operational objectives. It includes monitoring of operational execution effectiveness (activity D42), CS performance at both an individual commodity (activity D712) and integrated level (activity D212), and operational and CS objectives (activities D3121 and D411), as well as directing and implementing reconfiguration actions (activities D413 and D711). When CS performance begins to differ from desired levels, either because of degradation of CS capability or changes in operational objectives, reconfiguration actions will be triggered.

This type of closed-loop regulating ensures that the CS infrastructure is monitored and adjusted to maximize operational effectiveness during execution. The Air Force has emphasized flexible tailoring for force deployment, but it has made less progress in the ability to react quickly to changes once deployed. The feedback loop allows the flexible tailoring concept to be extended from initial deployment to employment and sustainment.

Establishment of Standing CS Organizational Nodes

The *TO-BE* architecture designates three standing (permanent) organizational nodes dedicated primarily to combat support: the theater-level Operations Support Center (OSC), individual commodity Inventory Control Points (ICPs), and a Global Integration Center (GIC). The rationale for standing organizations is two-fold. The first is to provide operational continuity and seamless incorporation of peacetime CS activities during transition to a contingency. Having a single node, such as the OSC, responsible for CS activities across the spectrum of operations optimizes time and energy during the transition to higher-intensity operations. It eliminates the need to transfer command responsibilities, minimizes the confusion and delay accompanying augmentation of wartime organizations, eliminates the ambiguity in redirecting information flows into wartime organizations, and leverages the peacetime knowledge base regarding regional and temporal infrastructure, transportation, and host nation idiosyncrasies. A standing OSC alleviates these shortcomings by using the same staff, organization structure, and information and communication networks in peace and war.

The second motivation for standing organizations is to provide uniform roles and responsibilities for a given organizational node in different theaters. Even if the organization occupying that node is theater dependent, the node structure ensures that the organization's role is well defined and corresponds to the organization occupying the same node in other theaters. This allows for intertheater consistency in activities and objectives, relationships with other organizational nodes, performance standards and metrics, and personnel training curricula. Such global consistency is critical for assessing and comparing the state of the CS infrastructure and readiness levels, arbitrating the allocation of resources between theaters, developing plans that may involve multiple theaters or intertheater interactions, and training and assigning personnel to staff the organizations.

Establishment of Centralized Management of CS Resources and Capability

The *TO-BE* architecture includes a GIC to monitor and integrate information regarding CS resources. Support requirements generated at the JFACC/AFFOR and OSC levels feed ICP plans, which are then centrally overseen and managed by the GIC and fed back into the JFACC/AFFOR and OSC. The GIC is thus able to provide a comprehensive CS perspective to the JFACC/AFFOR and OSC to help guide plan development and regulate activities during execution. It should have the visibility and clout to suggest alternatives where appropriate. In addition, it should have authoritative power to dictate production and acquisition rules to ICPs.

One of the primary goals of the GIC is to ensure that individual commodity support activities are coordinated to address total plan supportability. An integrated supportability assessment can then provide the input and feedback to the various planning steps to be sure that weapon system, airbase, and personnel resource requirements are accounted for and can be supported.

In the readiness phase, the GIC must manage support for both deliberate planning and training. In the crisis action planning phase, the GIC must analyze total weapon system sortie generation capability based on the individual ICP plans, as well as determine supply chain capability to sustain these plans (activities B21 and B22). These and other inputs contribute to a CS feasibility assessment. The GIC may need to suggest or impose adjustments to the support or alternative mission approach strategies. These monitoring and regulatory activities continue into the employment/sustainment phase, where the GIC is responsible for monitoring CS system performance, working with the ICPs to identify causes for system performance degradation and with the ICPs, OSC, and JFACC/AFFOR to design and implement get-well plans. With a single node for managing resources and capability, planners will have a reliable source of information and will be better able to develop informed, feasible plans.

Improved Ability to Monitor and Arbitrate Resources Across Competing Theaters

Because the GIC is able to monitor and analyze CS resource requirements and capabilities from a global perspective, a key responsibility is to monitor and arbitrate resource demands across competing theaters. This responsibility extends from readiness through employment/sustainment.

In the readiness phase, the GIC must integrate individual commodity CS plans for supporting both deliberate planning and training (activity A22 in Figure C.1). It would monitor weapon system readiness, adjust individual commodity support strategies to balance global resource demands, and arbitrate resources among competing plans.

In the crisis action planning phase, commodity support developed by the ICPs to support air campaign plans being generated by the OSC and JFACC/AFFOR often require the diversion of resources from other theaters. One of the GIC's primary responsibilities in this phase is to monitor the impact of any resource diversion on individual commodity and total weapon system readiness in other theaters (activity B21). Further, when such impact is deemed unacceptable, the GIC is responsible for working with the competing combatant commanders and OSCs, together with the ICPs, to make adjustments or develop alternative plans (activity B22). Similar monitoring and arbitration must occur during the employment/sustainment phase. The ability to monitor resource levels across competing theaters and to make arbitration decisions based on the new information increases operational capability where it is needed most.

RECOMMENDED USES

The process diagram and supporting database in this appendix offer the critical components of an operational CS execution planning and control architecture. The visual presentation and underlying content make it a valuable reference as the Air Force CS community transitions from the current architecture to the *TO-BE* concept. We next suggest how this material could be used to facilitate the transition.

Enhancing Air Force Doctrine and Policy on CS Execution Planning and Control

One of the shortfalls in the current system involves Air Force CS doctrine and policy. Because the CS execution planning and control concept is not well defined in doctrine, assignment of responsibilities to organizations is not well defined in policy. Proposed solutions include rewriting Air Force Doctrine Documents 2, 2-4, and 2-8 to address basic objectives and functions for combat support. New Air Force Instructions (AFIs) and possibly Tactics, Techniques, and Procedures (TTPs) documents should include the assignment of responsibilities, processes, and information flows to C2 organizations. To that end, the Figure C.1 diagram and database assign responsibilities, processes, and information flows to existing and new organizational nodes within the *TO-BE* system and can be used as a source document for rewriting existing doctrine and policy documents or developing new ones.

Training Material

In our site visits and interviews, we learned that many warfighting staff members are not adequately trained in their management roles. Most are assigned to regional CS roles from the wing level and have little or no experience with resource management at a regional level. Little formal training is available, leaving both operations and CS personnel to learn most of their responsibilities through on-the-job training. Solutions to the training shortfall were addressed in Chapter Five. One solution is to develop a CSC2 curriculum that incorporates CS execution planning and control into formal courses such as the Joint Aerospace C2 course, Air Force Institute of

Technology's (AFIT's) Logistics 399 and 499 courses, and Chief's Logistics Review's Logistics Officer Weapons School. The structure and content of the diagram and database could be translated into training material for curriculum development. The products uniquely reflect C2 activity across each phase of operations and at each echelon, and thus could be adapted for courses from SECDEF/Joint level to base level. Its HTML format lends itself well to development of web-based applications and training aids for distance learning and OJT. It could be further translated into a graphically oriented interactive product.

Another training shortfall solution calls for enhancing wargames and exercises with a higher level of CS fidelity. Products described here could be used to develop training and evaluation criteria, script events, and exercise the C2 nodes (e.g., a Total Asset Visibility database could be developed and integrated with the Global Transportation Network database to train logisticians on the global distribution system). Operations and CS personnel would have a more realistic training environment into which CS considerations were fully integrated.

Operations Requirements Documentation

This report identifies where both the *AS-IS* and the *TO-BE* CS systems would benefit from the enhancement or development of information systems and decision support tools. Because CS resources have been managed and funded by commodity with different organizations having commodity management responsibility, corresponding information systems have been implemented independently. The result is a myriad of stovepipe systems with little ability to share data or interface with other systems. Tools are needed to relate operational plans to CS requirements, convert CS resource levels to operational capability assessments, aggregate assessments to a theater or global scale, and to conduct tradeoff analyses of operational, support, and strategy options. Comprehensive operations requirements documentation is critical to the development of these types of tools. Before systems engineers can build the infrastructure and tools needed for the *TO-BE* CS capability, users must identify their requirements—what processes the tool is to facilitate, what information is to be captured or shared, and where the information must flow. The products discussed in this report are well suited as source material for requirements documentation for system architecture, decision support tools, and information system development. For example, the products would be useful in developing and maintaining the Air Force input into the JCS CINC 57 Category One Requirements for the Global Combat Support System (GCSS). They could be used by the AEF and C2 battlelabs to filter potential battlelab CS initiatives and by the Air Force Experimentation Office to help select tools for evaluation in the Joint Expeditionary Force Experiment (JEFX) series.

REFERENCES

DOCUMENTS

Aerospace Command and Control, Intelligence, Surveillance, and Reconnaissance Center, *USAF Command and Control CONOPs*, Vol. III, *Blue Order of Battle, Global Awareness for Expeditionary Aerospace Forces*, Langley Air Force Base, VA, 7 July 2000.

Department of Defense, *C4ISR Framework Document Version 2.0*, Washington D.C., 18 December 1997.

Feinberg, Amatzia, Hyman L. Shulman, Louis W. Miller, and Robert S. Tripp, *Supporting Expeditionary Aerospace Forces: Expanded Analysis of LANTIRN Options*, RAND, MR-1225-AF, 2000.

Feinberg, Amatzia, James Leftwich, Eric Peltz, Robert S. Tripp, Mahyar Amouzegar, and Russell Grunch, *Supporting Expeditionary Aerospace Forces: Lessons From the Air War Over Serbia*, RAND, MR-1263-AF, 2002.

Galway, Lionel, Robert S. Tripp, Timothy L. Ramey, and John Drew, *Supporting Expeditionary Aerospace Forces: New Agile Combat Support Postures*, RAND, MR-1075-AF, 2000.

Hanser, Lawrence M., Maren Leed, and Charles Robert Roll, *The Warfighting Capacity of Air Combat Command's Numbered Air Forces*, RAND, DB-297-AF, 2000.

Hillestad, R. J., *Dyna-METRIC: Dynamic Multi-Echelon Technique for Recoverable Item Control*, RAND, R-2785-AF, 1982.

Joint Chiefs of Staff, *DoD Dictionary of Military and Associated Terms*, Joint Pub 1-02, Washington D.C., 12 April 2001.

Peltz, Eric, Hyman L. Shulman, Robert S. Tripp, Timothy Ramey, and John Drew, *Supporting Expeditionary Aerospace Forces: An Analysis of F-15 Avionics Options*, RAND, MR-1174-AF, 2000.

Pyles, Raymond, *The Dyna-METRIC Readiness Assessment Model: Motivation, Capabilities, and Use*, RAND, R-2886-AF, 1984.

Pyles, Ray, and Robert S. Tripp, *Measuring and Managing Readiness: The Concept and Design of the Combat Support Capability Management System,* RAND, N-1840-AF, 1982.

Rumsfeld, Donald, *Defense Strategy Review,* 21 June 2001.

Rumsfeld, Donald, *Guidance and Terms of Reference for the 2001 Quadrennial Defense Review,* 22 June 2001.

Ryan, General Michael E., U.S. Air Force, "Aerospace Expeditionary Force: Better Use of Aerospace Power for the 21st Century," briefing, Washington, D.C., U.S. Air Force, 1998.

Tripp, Robert S., Lionel Galway, Timothy L. Ramey, Mahyar Amouzegar, and Eric Peltz, *Supporting Expeditionary Aerospace Forces: A Concept for Evolving the Agile Combat Support/Mobility System of the Future,* RAND, MR-1179-AF, 2000.

Tripp, Robert S., Lionel Galway, Paul Killingsworth, Eric Peltz, Timothy Ramey, and John Drew, *Supporting Expeditionary Aerospace Forces: An Integrated Strategic Agile Combat Support Planning Framework,* RAND, MR-1056-AF, 1999.

U.S. Air Force, *Air Force Basic Doctrine,* Air Force Doctrine Document 1, Washington, D.C., 1 September 1997.

U.S. Air Force, *Combat Support,* Air Force Doctrine Document 2-4, Washington D.C., October 2000.

U.S. Air Force, *Command and Control,* Air Force Doctrine Document 2-8 (Draft), Washington D.C., 16 February 2001.

U.S. Air Force, *Operation Plan and Concept Plan Development and Implementation,* Air Force Manual 10-401, Washington D.C., 1 May 1998.

U.S. Air Force, *Operational Procedures—Aerospace Operations Center,* Air Force Instruction 13-1AOC, Vol. III, Washington D.C., 1 June 1999.

U.S. Air Force, *Organization and Employment of Aerospace Power,* Air Force Doctrine Document 2, Washington D.C., 17 February 2001.

U.S. Air Force, *Readiness,* Air Force Policy Directive 10-2, Washington D.C., 1 Mar 1997.

U.S. Air Force, *Status of Resources and Training System,* Air Force Instruction 10-201, Washington D.C., 4 May 2000.

WEB SITES

U.S. Air Force, Joint Expeditionary Forces Experiment 99 Final Report, Air Force Experimentation Office website, https://jefxlink.langley.af.mil/, undated.

U.S. Air Force, Joint Expeditionary Forces Experiment 00 Final Report, Air Force Experimentation Office website, https://jefxlink.langley.af.mil, undated.

INTERVIEWS

ACC, ACC Regional Supply Squadron, transcript of interview, Langley Air Force Base, VA, 6 February 2001.

Headquarters 7AF, 7AF staff, transcript of email interviews, Osan Air Base, Korea, 10–11 December 2001.

Headquarters ACC, ACC Crisis Action Team and LG staff, transcript of interview, Langley Air Force Base, VA, 6 February 2001.

Headquarters AFMC, XPAO, transcript of interview, Wright-Patterson Air Force Base, Ohio, 21 February 2001.

Headquarters AFMC, Battle Staff, transcript of interview, Wright-Patterson Air Force Base, Ohio, 2 May 2001.

Headquarters CENTAF, AFFOR A-4, transcript of interview, Shaw Air Force Base, South Carolina, 8 February 2001.

Headquarters CENTAF, AFFOR A-3/A-5, transcript of interview, Shaw Air Force Base, South Carolina, 8 February 2001.

Headquarters, CENTAF AFFOR A-4, transcript of interview, Shaw Air Force Base, South Carolina, 25–26 April 2001.

Headquarters PACAF, 52 AOG and LG staff, transcript of interview, Hickam Air Force Base, Hawaii, 6–9 March 2001, 17–18 December 2001.

Headquarters USAFE, LG staff, transcript of interview, Ramstein Air Base, Germany, 4 April 2001, 22–24 January 2002.

Headquarters USAFE, 32 Air Operations Squadron, transcript of interview, Ramstein Air Base, Germany, 5 April 2001.

Headquarters Joint Chiefs of Staff, J-4, transcript of interview, Pentagon, Washington D.C., 23 February 2001.

Headquarters 13AF, Andersen Air Force Base, Guam, 13–14 December 2001.

PACAF, PACAF Regional Supply Squadron, transcript of interview, Hickam Air Force Base, Hawaii, 8 March 2001.

USAFE, USAFE Regional Supply Squadron, transcript of interview, Ramstein Air Force Base, Germany, 4 April 2001.